We Create Health One Truth At a Time

Praise for
We Create Health One Truth At a Time

"*We Create Health One Truth At a Time* is a work that is profoundly human, written wisely by a physician who has dedicated her life to curing at the root. Doctor Ana Negrón invites us to courageously question what we have accepted as a given in conventional medicine and with evidence and empathy reminds us that true prevention begins in the plate, in the community, and in caring for ourselves.

"As a clinical pharmacist who has transitioned to lifestyle medicine, this book resonated with the most profound in my practice and my story. The manner in which Dr. Negrón weaves science, clinical experience, and personal narrative is inspiring and profoundly transformative.

"This book does not only inform: it empowers. It is an essential guide for health professionals, patients, and communities that want to take control of their wellbeing and be part of a systemic change we so much need. Ana Negrón is an indispensable voice in the movement of health based in plants and equity within medical care. This book is a gift that sows conscience, compassion, and truth—one page at a time."

—Evelisse Capó, PharmD, DipACLM

"The most terrifying fear is to the unknown. Disease is feared because we don't even know how the human body functions: our life vehicle, our temple. Dr. Ana Negrón is definitely one of those doctors of the future, who strive to make the complexity of the body and even disease proper, simple; landing in a nourishment treatment based on scientific evidence, without covering up the atrocious deceits perpetrated by industries to which we have been subject to confuse us, disconnect us from ourselves, and avoid health at all cost. *We Create Health One Truth At a Time* is a masterpiece. All your doubts will be resolved and your fullness will shine once and for all."

—Dra. Marcela Sandoval-Garza, Board Certified Lifestyle Medicine Intensivist

"This book contributes an innovative vision from its introduction until the end... it will invite you to reflect, will touch your heart, awaken your mind, and will help you perceive a world full of possibilities."

—Evelyn Ortiz Avilés, Escritora/Presidente Instituto de Permacultura de Puerto Rico, Inc.

"...truly a wonderful compilation of sound science with relatable compassionate guidance which will serve so many! This is a GEM!"

—Saray Stancic, MD, FACLM

"This book is an inspiring guide for those looking to transform their mental and physical health through a more conscious lifestyle. With its profound focus on the whole wellbeing, healthful nourishment, and ecological respect, its pages invite a walk on the holistic path that wakes us to a whole life in harmony with ourselves and our surroundings.

"...a profoundly human, scientific, and educational book. We loved how Ana integrates knowledge with a narrative that feels near and sarcastic in the best sense—like that "ha!" that shakes you and makes you smile—something that resonates with the warm and direct style of our Latina culture.

"One of the virtues of this book is its capacity to explain physiology with creativity: theatric comparisons, like the ones used to talk about the microbiota, transform complex concepts onto living, poetic, and memorable scenes. It reminds us that science can also count on art.

"The analogies using scrabble and musical notes are masterful in explaining how the body does not distinguish the origin of amino acids. The importance is the science, not the myth.

"One of our favorite quotes is: 'Nourishment is too important for you to eat or stop eating something because of fear, before investigating the veracity of what you've heard somewhere.' It reflects the spirit of the book: empowers with evidence and respect for the truth.

"It also touches social topics with courage, such as the prejudices against vegans, and how this relates to the resistance of facing uncomfortable realities. It is not moral superiority, it is simply ethical and scientific coherence.

"We wouldn't just recommend this book, we would incorporate it for our students of Medicine and Nursing at the University, to help understand nourishment as a tool for health and for an empathic bridge with patients. For those of us who practice lifestyle and culinary medicine, it is a reaffirmation of the commitment we have made with health promotion, the rigorous study, and respect for life.

—Fabiola Flores Monsivais Médico Internista y Dr Chef Jorge Castillo Cirujano Maxilofacial Universidad Autónoma de Baja California

We Create Health One Truth At a Time

ANA M. NEGRÓN MD

Sunstone Press
Santa Fe

Sunstone books may be purchased for educational, business, or sales promotional use.
For information please write: Special Markets Department, Sunstone Press,
P.O. Box 2321, Santa Fe, New Mexico 87504-2321.
Printed on acid-free paper
♾

Library of Congress Cataloging-in-Publication Data

Names: Negrón, Ana M., 1942- author
Title: We create health one truth at a time / Ana M. Negrón MD.
Description: Santa Fe : Sunstone Press, [2026] | Includes bibliographical references and index. | Summary: "Ana M. Negrón, MD, explains how our human body is put together, what nourishes it, and lays out the best ways to maintain it, including practical suggestions and recipe templates, with an emphasis on reevaluating the widespread dependence on pharmaceuticals and supplements and basing one's health on facts and truth"-- Provided by publisher.
Identifiers: LCCN 2026020122 | ISBN 9781632937940 paperback | ISBN 9781632937957 hardback
Subjects: LCSH: Human body | Nutrition | Diseases
Classification: LCC QP38 .N385 2026 | DDC 612--dc23/eng/20260423
LC record available at https://lccn.loc.gov/2026020122

WWW.SUNSTONEPRESS.COM
SUNSTONE PRESS / POST OFFICE BOX 2321 / SANTA FE, NM 87504-2321 /USA
(505) 988-4418

Dedication

To you curious reader, who insists on knowing the truth.

CONTENTS

Foreword, LeAnne Campbell, PhD / 15
Preface, Ana M. Negrón / 17
Acknowledgments / 21
Introduction, Ana M. Negrón / 23

PART ONE / 29
Know Your Body

Your Body Is Unique / 31
Tentative Beginnings / 32
Construction Materials / 33
Totally Dependent Genes / 34
Epigenetics and the Exposome / 35
Consequences of Not Paying Attention / 36
Nourishment and Health Go Hand in Hand / 36
Range of Responses / 37
Organs and Their Systems / 38
Twenty Feet of Intestines / 38
The Mouth / 39
The Esophagus / 40
Next Stop: the Stomach / 41
The Duodenum / 41
The Pancreas / 43
The Liver / 45
The Microbiota / 46
The Intestinal Lining and Mucosa / 48

The Kidneys / 50
The Uterus / 51
The Architecture of the Skeleton / 52
Muscle Fibers and How They Grow / 53
The Brain / 54
Sleep and Dreams / 55
Who Am I? / 57
The Lungs / 57
The Lymphatic System / 58
The Cardiovascular System / 58
Adipocytes / 62

PART TWO / 63
Food and Its Nutrients

Carbohydrates / 65
Fiber / 66
Protein / 67
Scientific Curiosity / 70
The Protein Cult / 70
The Proteic Dictionary / 73
The Word Dictionary / 74
Lectins / 76
Fats / 77
DHA / 78
Vitamin B_{12} / 79
Calcium / 80
Cow's Milk / 80
Money Leaves a Trail / 84
Animal Meat / 85
Veganism / 86
Eggs / 87
Oil / 88
Your Reputation Follows You / 88
A Balanced Diet / 90
Calories / 90
Concentrated Energy / 91

The Horse / 92
The Human Body / 93
Calories Courtesy of the Food Industry / 93
Whole and Processed Food / 94
Factory Food / 96
Our Biological Pattern, Scheme, or Wiring / 97
The Pleasure Trap by D. Lisle and A. Goldhamer / 99
Traditions and Habits / 101
Free Human Beings Are Unpredictable / 103
Eyes of an Eagle and Eyes of a Mouse / 104
The Papaya / 104
Supplements / 104
The Faces of Reductionism / 106

PART THREE / 109
Diseases of the Day

Obesity / 112
Natural Scarcity and Fabricated Scarcity / 117
Diabetes / 120
Reverse Insulin Resistance / 121
Diabetes and Insulin Resistance / 122
Hypertension and the Kidneys / 124
Coronary Disease and Cholesterol / 126
Hypertension and the Heart / 128
Hypertension and Erectile Dysfunction / 131
Hypertension and the Brain / 131
Menopause Is Not a Disease / 133
Osteoporosis and Bone Fragility / 134
Arthritis / 136
Sleep Apnea / 137
Cancer / 137
Autoimmune Diseases / 139
Chronic Inflammation and Healing / 140
Laboratory Tests / 143

PART FOUR / 145
General Topics of the Health Care System

Pharma Deserves a Section / 147
Health Is Free, Sickness Is Lucrative / 147
Dear Colleague / 149
The Management of Disease / 153
The Public Has a Right to Know / 154
Conflicts of Interest / 154
Public Health Medicine / 156
Lifestyle Medicine / 158
Medical Education / 159
Drugs as Third Cause of Death / 160
Dissonances / 161
There Is One Diet / 162
We Inject Nutrition Into Medicine / 166
Healers / 166
Science Topples and Dethrones Myths / 167
Wrong Information and Disinformation / 168
How Is a Vaccine Made? My Supersimplified Version / 175
The Recipe for Optimal Nutrition / 178
The Blue Zones: Mirrors of Longevity / 178
Subsidies to the Cattle Industry and Its Significance / 180
One Step Forward and Two Steps Back / 183
The Current USDA Plate / 183
Supplements for Prevention / 185
Supplements are Isolated Fractions / 186
Model for Optimal Nutrition / 186
Truth or Fiction in Nutrition / 189

PART FIVE / 193
How I Landed Here

My Trajectory—Sort Of / 195
Beginnings / 195
Eyes Open to the Ocean / 197
Fifteen Hundred Miles / 198

1957 Set My Path / 199
In School / 200
At Home / 200
One World Closed and Another Opened / 201
In My Practice / 201
A Sabbatical / 202
Living on Pharmaceuticals Is Unsustainable / 204
The Community and I / 205
Greens on a Budget—From 2003 to 2023 / 207
Fresh2You / 209
Zoom / 210
Twenty Feet of Intestine / 210
Paper and Pencil / 211
Curiosity / 212
Latino Tradition and Culture / 213
My Dream: Latin America—One Enormous Blue Zone / 215
Stories From the Community / 216
Love the Truth / 217
About Freedom / 217
Use Your Power / 218
Look in the Mirror / 218
Among the Lettuces, One Cabbage / 220
Anecdotes That Uplift and Inspire Me to Keep Going / 221
Reflections / 228
Logic and Dissonance / 229
We Can Help Pave the Way / 229
More Anecdotes / 231
Glossary / 239
Living the Change / 240
It Is Not Easy / 241
The Saturday Breakfasts / 242

PART SIX / 245
Science Is Not Neutral or Balanced, Science Is Objective

Where to Start / 247
What Is Your Understanding of Health? / 247

Freedom / 248
How to Stock the Kitchen: An Example / 249
Habits / 250
New Habits / 251
Simplify Your Routine / 251
What Will You Eat? / 251
Social Interactions / 23
It Is Never Too Late / 254
The Virtual Kitchen Tour / 256
The Kitchen Environment / 258
Home Composting Made Simple / 259
Why Should We Not Throw Organic Matter Onto the Landfill? / 259
What Should We Compost and How? / 259
What Can We Add to the Compost Pile or Bin? / 259
What Should Not Be Thrown Into the Compost Bin? /260
From the Kitchen to the Compost Bin / 260
Balance / 260
Tips and Solutions / 260
Benefits of Composting / 261

Epilogue / 263
One Last Task / 263
Sample Recipes / 265
My Human Body / 273
Index / 275

Foreword

Never before has there been such a critical need for Dr. Negrón's book, *We Create Health One Truth At a Time*. Diseases such as diabetes, cancer, heart disease, autoimmune disease, hypertension, Alzheimer's, and obesity are now affecting populations across the globe, even reaching villages in remote corners of the world. More and more people are dying from these preventable diseases, suffering tremendous pain, and having to navigate astronomically high medical bills.

In La Cumbre, Dominican Republic, a small mountain village of 40 households where I live most of the year, I see this up close. I first arrived in this area over 35 years ago, when I was stationed to live and work in a rural health clinic. My neighbors and other village health workers who were alive at that time remember, as well as I do, a time when people checked into the clinic with infectious diseases, such as pneumonia, hepatitis, tuberculosis, and a wide array of parasitic infections. Rarely did we see people with diabetes, heart disease, or cancer. But today it's different. Obesity is rampant, as well as diabetes, heart disease, hypertension, and cancer.

This month alone, we had multiple funeral services in La Cumbre as well as one-year prayer services for those who died a year ago. Níne was 59 years of age, a grandmother of eight, and a mother of four. She had her leg removed two months before her last dialysis treatment. A month later, she passed away. Cause of death: diabetes. Beethoven was 28 years old. He had severe stomach pain and was not able to eat. He went to the clinic and was told he had advanced diabetes. They began dialysis. Shortly after this, his kidneys failed, and as his mother said, "They dried up." He died shortly after this. Cause of death: complications centered on diabetes. Elena, a mother of two and grandmother of four, was taking multiple medications to deal with high blood pressure, dizziness, high cholesterol, and indigestion. She had a heart attack at 42 years of age. Cause of death: heart disease.

Why are we witnessing and seeing more of these diseases today? What can we do about it?

Dr. Negrón's *We Create Health One Truth At a Time* provides answers and guidance for not only surviving but also thriving despite these disturbing trends in global health. She teaches us about our miraculous bodies, how our organs function, what they need, and the ways that food either nourishes or harms us. She shares the research and provides guidelines for how we can critically assess the science. She invites us to understand our choices, our decision-making processes, and our ability to take control. As you read, you'll understand; these teachings are powerful and transformative. Read it, savor it, reflect on it, take notes, dive into the references, reread it, and embrace it. Plan how you can create health for yourself, your family, friends, and neighbors. It will enhance—and may even save—your life!

—LeAnne Campbell, PhD, President T. Colin Campbell Center for Nutrition Studies

Preface

Truth is not neutral or balanced; truth is objective.

We own a marvel of kidneys, brain, ovaries, testicles, intestines, liver, thyroid, pancreas, adrenals, pituitary, and much more, encased in a waterproof skin. Inside, it has miles of elastic tubing through which liquid gold is pushed and circulates with each beat of the heart.

As arteries come out of the heart, they branch out all through the body until they are thread thin capillaries that then connect to make ever larger veins that bring blood back to the heart. At the level of the capillaries there is a unique exchange: the blood delivers oxygen and nutrients to each cell and picks up the waste that will be eliminated in urine and stools plus the carbon dioxide that we give out with each sigh.

The body is also connected by a network that sends and receives messages to the brain along nerves wrapped in insulation, like the cord of a lamp, to prevent the electricity from dispersing.

Our curious minds are capable of understanding all this.

We create health with exercise and rest, free of toxins, excellent nourishment, without undue stress, living with purpose and guided by love. Enjoying good health is much more than not being sick.

The purpose of this project is to dismantle false notions about our bodies and nutrition and to replace them with verified truths. If we are to make good decisions, we must trust the information we use.

Going forward, we will explore the organism, some of its functions, and how food is able to support it. We shall stop along the way that brought me to this point and share stories. Since there are many whole plant recipes available that don't use flour, sugar, or oil, you will only find some templates that can be amended to suit your taste, put in practice what you learn.

This project pretends to encourage and support a world of all and for all; recognizing this is an endless task and we must all participate.

The written word is incapable of conveying everything, so enjoy the images, photos, graphs, literary and political references, read between the lines, and give flight to your imagination. References from the medical literature are simply to guide your curiosity.

The various medical conditions mentioned are connected to our chosen diet and lifestyle and the great power we have to reduce suffering. Nonetheless, no matter how well we eat or live, no one is immune to disease. It is essential and we deserve a medical system based on scientific investigation free of economic interests. Any question that you have regarding a personal condition should be discussed with a health professional who shows respect for the truth.

NUTRITION AND HEALTH

Scientific investigations have been uncovering more and more mechanisms of the indisputable relationship between nutrition and health. Based on these discoveries, we validate those past understandings that are correct and discard those that prove to be false. It is very important to understand that scientific investigations do not have a preordained objective, rather they go seeking truths. Scientific investigations yield surprising results (for example that the intestinal microbes turn fiber into important and nourishing short chain fatty acids), or confirm suspicions (that tobacco causes lung and bladder cancer), or they shake us to the core (as we will see with the investigations on animal protein).

For me, studying has been a discipline and a pastime. I trust that as you discover the collaboration between the body and our environment, your admiration, respect, and love for the Earth of which we are a part will grow, and so will your desire to protect her.

THIS BOOK IS FOR YOU

If you depend on pharmaceuticals, supplements, or processed foods, I wrote this book for you.

If you swapped oranges for orange drinks, tortillas for pizza, mushrooms for nuggets, and potatoes for chips, it is also for you.

If you abandoned rice and beans thinking this is food for poor people, it is my gift to you.

If you seriously want to recover health, I hand crafted this book especially for you.

If you have been part of the medical-pharmaceutical complex, put your name next to mine and spread its content.

Acknowledgements

You will never know that with a gentle word, a brief gesture, an apparently insignificant smile, you helped propel this project; that your courage accompanied me through long stretches, distances that I wouldn't have been able to traverse alone; that your vision sharpened my senses so I could hear voices calling from a future we have yet to know and that I hope exists. To all of you, thank you.

I was initiated early in the rigorous manner of studying and exposing scientific work with people of supreme integrity. John Robbins opened my heart to the evident cruelty of using animals. Antonia Demas showed me that to learn we must engage all our senses. T. Colin Campbell lights my path.

Saray Stancic shook me with her scrupulous story of what is missing from medicine. Fabiola Monsivais and Jorge Castillo gave voice and eloquent application to investigations that I had only read in English; I feel immense pride knowing their work which continues to empower me for the labor I have undertaken during the past twenty years with immigrants from Latin America; in the precious moments before closing this document, their keen insights left a palpable trace. I am grateful for the Spanish I learned as a child and that I proudly use to communicate this book in its original version.

Evelisse Capó, colleague with whom I have walked for years, as daughter and ambassador of nutrition based on plants, known throughout all Latin America, obtains her inexhaustible energy from a disciplined care so replete with joy that is a stimulus and a support. Marcela Sandoval took me to another plane with her warm professional empathy directed toward our bodies.

Grateful to know Myriam Masihy, a generous and effective investigative reporter who takes this information in a precise and clear form to the Latino community. I had the fortune of receiving the professional comments of Zinnia Cintrón-Marrero about written language conventions.

I couldn't have developed a fraction of what I share with you in this project if it wasn't for the community, my students, and my patients who show me how they learn, how they stumble, how they retain information, how they use it, how they share it. I am forever grateful to their inexorable insistence, their unending questioning, their unnerving reluctance—Above all I celebrate their magnificent successes.

As Violeta Parra wrote, "Gracias a la Vida," thanks to life, for it has given me so much. To my son and composer Jason Binnick who has kept me anchored in reality and has supported my imagination. To all my fellow travelers with whom I hope to keep sharing the way.

Finally, I have the privilege of counting on my editor James Clois Smith, Jr. at Sunstone Press, who in spite of not commanding the Spanish language placed his confidence in the original version of this project and treated it with the respect and care that I wish every author could receive.

Introduction

Good health is a state of total physical, mental, and social wellbeing that facilitates all of our functions.

A vast amount of the information about food that we have today comes from studies that were designed and financed by the same food industries, whose purpose is not to discover truths, but to favor the sale of their products.

REVOLVING DOOR

Many investigators held lower paid positions in independent academic institutions before they transferred to the more lucrative private world. After collaborating with pharmaceutical industries, food companies, and supplement businesses some return to government departments, where they help craft nutrition guidelines with a commercial bias.

For example, the Women, Infants and Children (WIC) program distributes cheese to their mothers with instructions of feeding it to their children to assure bone health. Did you know that these recommendations came without merit from the National Dairy Council?

> Reference:
> Marion Nestle, professor at NYU and author of *Food Politics* and many other important writings about the economy and industry, published on her webpage 168 studies financed by the food industry; 156 of those arrive at conclusions that favor said industry.

INFOXICATION

The term "infoxicación" in Spanish was coined by Alfons Cornella and refers to intoxication by information and the current flood of information which can be so disorienting. The sheer quantity of messages published annually about food is staggering. So, what can we do to keep from drowning in this sea of data?

Well, we could consult the *Nutrition Guidelines*. The US Department of Agriculture (USDA) updates this national guide every five years. It used to be represented in a pyramid and since 2010 we know it as a color-coded plate (MyPlate.gov) All federal or private entities that serve food to the public follow this guide. There is a similar one in your country.

Here is the problem, the USDA is charged with a double mission: support the cattle industry and publish nutrition guidelines for the public, which means that they operate under a conflict of interests. On one hand they advise us to eat less saturated fat, but it grants subsidies to the meat and dairy industry whose products are the most important vehicle for saturated fat. How could we believe that their recommendations are impartial?

INVESTIGATION

What if we call upon the National Institutes of Health (NIH)? They publish scientific investigations about food, don't they? For starters, not one of the twenty-seven institutes that make up the NIH is dedicated exclusively to nutrition. Additionally, most of the studies approved by the NIH are to demonstrate the reductionistic effect of one nutrient or drug upon some biochemical reaction. On top of that, 67% of its funded studies are performed on animals. Could we infer that the results or conclusions of such studies apply to the human as a whole or even to the human being at all?

Okay, what if we do a search of all the investigations published about nutrition in humans? There are mountains of published studies, but in the last fifty years most were commissioned, designed, or realized by the same industries. These studies are carried out not to discover truths, but to promote sales. So, their resulting recommendations are tied to financial interests: cow's milk is the primary source of calcium, we need animal protein to develop muscles, athletes must consume eggs, fish oil is heart healthy, we better rely on supplements, and so on. You might have even heard this propaganda from a physician. Do not be surprised, since the prescribed continuing medical education is subsidized in particular by the pharmaceutical industry.

It should be noted that independent investigators who labor with integrity mainly in academic silos, do not disseminate their findings as earnestly as the for profit industries nor do they use accessible language in speaking to the public. In fact, many academic results don't even make it to health professionals. Academicians keep their heads buried in their investigations paying little attention to the unfettered propaganda and leaving the path clear for commercial interests to paint the panorama in their favor: that carbohydrates make you fat, that without animal protein we are toast, that coconut oil is the best. Open any medical journal or popular magazine and you will see these advertisements pretending to be science.

Reference:
American Diabetes Association: corporate sponsors

The constant bombardment of propaganda promotes a reductionistic view of food and health. It makes us believe that making decisions about nutrition is something complicated and out of our reach, that we must know about zinc and magnesium (both are abundant in plants), that we must worry about the omegas (also present in plants), how about carnitine (the body produces this amine in liver and kidneys and its excess is implicated in the production of the troublesome trimethylamine), etc. Afraid of ending with a nutrient deficiency, many worried well people decide to top off their diet with supplements. How could we criticize them? The problem is that physicians also fall for this.

The good news is that there are many scientific studies published half a century ago, before the current obscene influence from industries.

Plus, a growing number of independent peer reviewed clinical studies published in reputable medical journals. However, they are buried under an avalanche of paid disinformation.

SCIENCE IS OBJECTIVE; IT IS NEITHER NEUTRAL NOR BALANCED

Science seeks the truth, no matter how inconvenient. Science is objective, it does not deny some findings to favor others. It checks results many times. It does not pay for its studies with earmarked money, as the for profit industry does. Independent science does not use propaganda to flood the market with prejudice and bias. It does not spend millions (does not have them) advertising on the internet. Nor does it buy giant ads on interstate highways. True science is not glamorous and does not invest in influencers (should it?). Their reports are austere and rarely invite reading, so you are unlikely to learn of the results. He who has the most money pays for the brightest lights.

Check out the following study: it pretends to establish that "dairy products are a valuable source of calcium and other nutrients that benefit bones and are therefore considered important for growing bones." Its authors are employed by the Yili Group, a dairy company. Always investigate the source, before accepting any information or recommendation.

> Reference:
> The Effects of Dairy Product Supplementation on Bone Health Indices in Children Aged 3 to 18 Years: A Meta-Analysis of Randomized Controlled Trials. *Advances in Nutrition* 2023 July 4;14(5):1187–1196

THE LATINO CULTURE IS CONSIDERED HEALTHY

Our Latino community enjoys healthy traditions known throughout the western world. Nonetheless, it too has succumbed to myths such as: rice makes you gain weight, tortilla contributes to diabetes, beans are poor people food, meat is indispensable, milk grows children big and fat (this last one is true). These ideas have colonized our brains, they have distanced us from the land and have tried to displace our traditions. The worst part is that they have caused us heart disease, obesity, diabetes, and cancer.

We throw away our hard earned income on meat, eggs, cheese, oils, and supplements (for good measure). Refreshments and pastries are our daily bread. We revere energy drinks. As a result, younger and younger people are letting their health slip away.

We seek more protein that we need, believing it gives us energy. We look for cures in ketogenic diets (low in carbohydrates and high in fat), in appetite suppressing drugs, and in commercial shakes. We spend on supplements, supposed antioxidants, and processed products commonly misrepresented as "natural." Influencers from the world of theatre, the arts, sports, and music speak of these products with fervor. Don't let them fool you.

Health is not inevitable, but neither is disease. If we regularly eat enough nourishing food, we can aspire to live in good health. Think of how many one-hundred-year-old people living without disabilities there could be. We will speak of the Blue Zones later.

What we insist in avoiding will always follow us, what we face will keep us company.

Throughout the world, many people practice some type of religious fasting at least once a year. Typically, they abstain from sweets and animal products for twenty-one to thirty days, during which time they report feeling super well physically and spiritually. They even lose weight, even though these fasts are not considered a diet.

Imagine if during such fasting periods people would wake up to the true price of following an animal and processed food diet; more would continue some version of the fast and see lasting benefits. Churches should play a more active role in raising this consciousness.

If you are looking to recover health, consider that the power might be in your grocery shopping list and in what you keep in your refrigerator and cupboards. Don't wait for permission to stop inflaming your body with meat, chicken, milk, eggs, cheese, pastries, and fritters. You don't need health insurance to start nourishing your body with more beans, rice, tortillas, cabbage, mango, Swiss chard, sweet potato, and cauliflower. And you certainly don't need the blessings of a health professional who lacks nutrition education to eat more lentils, mushrooms, broccoli, quinoa, blueberries, and pumpkins.

However, you do need information that is correct and then you must put it in practice. The following documentary is educational and inspiring, and you will see me touring a community garden and also cooking with people. Check it out and enjoy it.

> Reference:
> *Comida como cura (Food As Cure)* Documentary in Telemundo Miami realized by Myriam Masihy

PART ONE

Know Your Body

Our body is an organism of about 37 trillion cells (37×10^{12}). Each cell is enveloped in a layered membrane composed of protein and fat. Most of the proteins in the body are an integral part of its structure and frame (bones, muscles, tendon, and organs). The rest of the proteins travel in the blood and are involved in the thousand and one activities of the organism as vitamins, enzymes, and hormones.

YOUR BODY IS UNIQUE

The DNA (deoxyribonucleic acid) codifies our being with a user's manual for life. It directs the specialization of the cells; how to make blood, hormones, and every organ. It is a map that guides our body on how to live, without our conscious knowledge.

It also brings instructions on how to adjust when conditions are suboptimal. Some of these adjustments are noisy and call our attention with pain, fever, weakness, swelling, and other pesky symptoms. Many others adjusments to suboptimal conditions are silent, such as high blood pressure, elevated cholesterol, fatty liver, or a cancerous mutation. Suppressing symptoms or tolerating poor eating and lifestyle practices for long, interferes with the natural functions and activities of the body and could end up in disease.

The body works twenty-four hours a day seven days a week. It will do this for the better part of a century carrying out countless functions. Our job as owners of this world marvel is to maintain the best conditions for its 37 trillion cells to function and prosper.

And yet, who hasn't abused it? This Hawaiian prayer (Ho'Oponopono) is a humble reminder of the relationship we have with our loyal companion. I learned it from my friend and colleague Doctor Marcela Sandoval, author of *"Evidente: La medicina del respeto"* [Evident: The Medicine of Respect]. Add your own personal touch when you recite it. Remember that you are speaking to your body.

Ho'Oponopono:

Forgive me
I am sorry
Thank you
I love you

TENTATIVE BEGINNINGS

For a human being to have a start, an egg must meet a random sperm and form a zygote. Upon meeting, each of these two cells contributes half of the genetic material of the new organism. The union does not always guarantee an embryo; 60% of zygotes are not viable and die without our awareness. Imagine a factory where some parts are defective and must be discarded before completing the final product.

The ovum brings enough nourishment to potentiate the division of the new cell for several days. The sperm doesn't even carry a toothbrush. A week into the divisions, the growing morula will try to anchor itself inside the uterine cavity. If successful, it will receive energy and nutrients from the mother, while each new cell turns into a tireless factory.

Half of this now hollow sphere will become the embryo and the other half will develop into the organ we know as the placenta. How does it know to do this? For the nine months of pregnancy the placenta insures constant communication between fetus and mother by flowing nutrients and oxygen in one direction while waste and carbon dioxide flow the opposite way. This new entity or embryo is totally depends on the mother ship. Mother and fetus will live connected through the placenta until this one is discarded after delivery.

The growth and development of the embryo depend on the materials it receives through the placenta. The embryo, later a fetus, obtains the macro and micro nutrients exclusively from what the mother eats. The moment we can exist independently outside the uterus, we are a person. Until then, we are totally dependent on the mother.

CONSTRUCTION MATERIALS

The formation of each new cell requires proteins and fats plus water and numerous micronutrients, minerals, vitamins, and enzymes. It is also necessary to have a constant infusion of energy mostly in the form of glucose which is derived from carbohydrates.

Coincidentally, all this is the stuff of what plants are made. We could obtain fats and proteins from meat, eggs, or even junk food, but we would be lacking in rich carbohydrates and the many nutrients that are found only in broccoli, cauliflower, beets, sweet potato, tomato, pepper, pumpkin, pea, garbanzo, lentil, soybean, rice, wheat, corn, oat, quinoa, mango, apple, papaya, pineapple, mushroom, almond, coconut, pistachio, olives, peanut, and avocado... We are better made from plants.

TOTALLY DEPENDENT

When we finally come out into the world, we still depend on our mother. For one year, her milk will be our perfect food. Every female mammal elaborates mother's milk for her young. A human baby born of about seven pounds will weigh about 20 pounds on their first birthday. Meanwhile, a calf that weighs 60 pounds at birth will turn into a 700-pound yearling. It makes total sense that the composition of human milk should be different from the milk of cows, mares, rats, or whales.

For example, the amount of calories, protein, and calcium in eight ounces of human milk is less than in eight ounces of cow's milk, yet it is exactly what a baby of the human species requires to grow at their rate. Assigning a higher nutritive value to cow's milk is a mistake which only serves the dairy industry, since they must sell their product to get government subsidies.

Babies nourished exclusively with breast milk obtain most of their energy from carbohydrates: lactose, glucose, and fructose (45% of the total calories). The protein in mother's milk is about 5% to 8% of the total calories and provides all the amino acids that a baby needs to grow and prosper. The fatty acids (50% of the calories in human milk) are exceedingly important for its metabolism as well as for the development of the immune and nervous systems. As the baby grows, the fat requirements decrease in favor of more carbohydrates.

GENES

In the nucleus of each cell we find the 23 pairs of chromosomes that we received from father and mother (46 in total). They hold our genetics, our inherited characteristics.

Each chromosome, made of DNA (deoxyribonucleic acid) molecules, contains between hundreds and thousands of genes responsible for constructing the human body. When we speak of a gene, we are referring to a DNA interval capable of translating into a functional protein.

It is curious that the genes responsible for the 100,000 proteins the body requires occupy only about 2% of our DNA. What would be the function of the remaining 98%? It so happens that this 98% is incredibly important, since it regulates where and when we are to express or manifest the 2%. In other words, the genes are subject to an immensity of instructions and permissions.

The following diseases are genetic, meaning that they are inherited. Perhaps you have heard of some:

> Arrhythmogenic right ventricular dysplasia/cardiomyopathy, Brugada syndrome, Charcot-Marie-Tooth disease, Cleft lip and palate, Cleidocranial dysplasia, Cystic fibrosis, Familial adenomatous polyposis, Hirschsprung disease, Huntington disease, Klinefelter syndrome, Kniest syndrome, Marfan syndrome, Mucopolysaccharidosis, Muscular dystrophy, Sickle cell anemia, von Hippell-Lindau disease.

On the other hand, heart disease, diabetes, cancer, Alzheimer's, multiple sclerosis, rheumatoid arthritis, and many other common conditions, which you undoubtedly know or might even have, are not on the list because while you may well inherit the susceptibility, they require an epigenetic push to manifest themselves.

EPIGENETICS AND THE EXPOSOME

If you inherit the genes for Huntington's disease, they don't need a trigger to manifest. Likewise, if you inherit the genes for cystic fibrosis, it means you have the disease. However, you could inherit deficiencies for metabolizing alcohol, but if you never drink you will not develop alcohol use disorder. Similarly, you could inherit the susceptibility to gluten sensitivity, but if you never consume such protein (from wheat, barley, or rye) you will live free of celiac disease or sprue.

Food is very important in epigenetics, but it is by no means the only factor that acts upon our genes. Hence, we speak of the exposome: the great many factors or cumulus that act upon the individual and whose effects contribute to our personal and generational state of health.

The exposome refers to negative environmental forces, such as contamination, exploitation, abuse, violence, activities that destroy our planet, poverty, homophobia, racism, sexism, and all the -isms. The exposome also includes positive forces like love, support, collaboration, education, rest, good nutrition, freedom, and everything that sustains and protects a community.

It means that by controlling certain factors, we can control which genes are expressed. This is quite heartening! This also means that we should not blame our genes if with our diet and lifestyle we are turning on diseases.

To surround ourselves with a mostly positive exposome, we need clean nutrition, daily exercise, enough hours of restorative sleep, a community, life with purpose, and support.

CONSEQUENCES OF NOT PAYING ATTENTION

The poorer we are in body, spirit, and environment, the greater the risk of developing and manifesting chronic non contagious diseases. It is evident that an epidemic of chronic diseases without a viral or bacterial agent has been killing us. While the exposome identifies which environmental factors contribute to the expression of our genes, epigenetics is the study of how food and the environment affect the expression of our genes.

The aforementioned level of lifelong stress, oppression, persecution, and poverty are extraordinarily important factors that translate into the type of proteins that we produce. Investigations demonstrate that persistent levels of stress end up as genetic changes that get passed on to the next generation.

This project will focus on food. However, numerous works show that the intolerable life circumstances we have created for too many in the population, erode their capacity to live full and healthy lives.

NOURISHMENT AND HEALTH GO HAND IN HAND

It is somewhat easy to identify what gave you a stomach ache, diarrhea, vomiting, or an allergic rash because it is temporally related. However, it is more difficult to associate acid reflux, menstrual cramps, colon cancer, or the fatigue you have been suffering for months with your dietary pattern. Our poor health is irrefutably associated with the western dietary patterns, a sedentary lifestyle, and toxic stress, yet we have decided to ignore it.

Only with the proverbial retro-scope (a lens that affords us the benefit of looking back) can we see how unwise it was to defend our right to eat donuts, pork chops, sausage, and bacon. How many swim in cheese, chips, and fried chicken for years before they are one day surprised by chest pain, the thirst of high sugar, or a malignant mass.

The weaker we allow our bodies, spirit, and environment to get, the worse are our defenses and the greater the probability of developing and manifesting all types of diseases.

Spoiler alert: I don't mean to jump ahead, but the scientific evidence seems to indicate that to achieve a balance with all the other pillars of health, we must optimize our nutrition by avoiding animal protein and consuming a great variety of whole plants every day.

RANGE OF RESPONSES

As we stated earlier, very few diseases are inherited strictly through genes, but we all inherit genetic strengths and weaknesses from our mother and father. The hook is that we don't discover these until they are challenged.

Our genes will express in favor, neutral, or negatively depending on the balance among these inherited tendencies and according to the conditions and stresses to which we submit them. Logically, some conditions are outside of our control, such as living in an occupied territory, in a time of war, under oppressive laws, or in a contaminated environment, but many others are in our hands.

For example, if you have inherited poor genetic tools to metabolize nicotine and other carcinogens in tobacco, your risk for developing lung cancer is greater if you decide to smoke. Let's say instead that you inherited the likelihood to be harmed by alcohol, but you never drink? You have the power to break the chain of disease that runs in your family through the consumption of alcohol.

Perhaps your genes tend to deposit saturated fat in liver and muscle cells which prefer to function lean, but you follow a diet based on whole plants low in fat. You therefore reduce your risk of developing Type 2 diabetes to almost zero. In a similar way, if you do not burden your organism with animal protein, you could avoid the cholesterol plaques that have killed so many of your relatives. It is evident that the western diet and lifestyle are putting our genetic weaknesses to the test at an ever younger age and more extensively.

The genes that we inherited are a unique combination never before seen, like fingerprints. What we can say about them is that they have allowed some diseases to manifest in our families and that perhaps we can break the chain. How smart would we be if we decided to protect our genes in advance?

ORGANS AND THEIR SYSTEMS

In this section we will explore parts of the body to decide how best to support it. Let us begin with the intestine, since that is how the construction and maintenance materials enter the body. Without them there is no life.

TWENTY FEET OF INTESTINES

Plants make nutrients which they store in their roots, skin, pulp, stem, leaves, fruit, and seeds. Their bodies are a marvelous web of multicolor carbohydrates (digestible and indigestible or resistant), proteins, fats, micronutrients, minerals, phytochemicals, and water. As soon as a bite enters the mouth and we start chewing, digestive enzymes begin to break the chemical bonds between the molecules that make up the food.

Whole plants are a treasure of macro and micronutrients in a matrix of fiber (non-digestible carbohydrates) bathed in water, phytochemicals, vitamins, minerals, and an untold amount of antioxidants. We free them as we destroy the fiber cage, so to speak. We break down food by hand, with a knife, or grinding it with our teeth. Other ways that facilitate the liberation of nutrients include boiling, baking, blending, or freezing, which physically breaks the fiber that keeps the nutrients trapped.

More than 60% of the weight of vegetables is water trapped in its fiber structure, which contributes to the firm and crunchy consistency of fresh broccoli. Fiber is inelastic. When water freezes, its molecular structure takes up more room, which stretches and breaks the fiber cage. Once thawed, the frame now broken cannot remain turgid and the broccoli turns flaccid. Have you noticed how water "grows" in the ice cube tray? Did you ever freeze soup in a glass jar to later find it broken? That's because the molecules of frozen water take up more space.

Throughout digestion, the fiber matrix continues to break down releasing its nutritional content to the intestine. This paced and measured process facilitates the availability of nutrients and water at the optimal rhythm for the body to use them. Imagine on the other hand a smoothie of fruits and vegetables. Since everything is already turned loose, your body has no control over the speed at which the nutrients are discharged. This avalanche presents a challenge to the intestinal tract unless you sip the drink very slowly over a couple of hours. Consider using smoothies as an occasional refreshing treat and not as a substitute to chewing your food.

THE MOUTH

The digestive system begins in the mouth. Its warm and humid environment is almost always full of nutrients that feed more than 700 species of bacteria, fungi, virus, and protozoa—most of them beneficial. Saliva with antibacterial and antiviral properties bathes the mouth defending us against pathogenic (disease causing) microorganisms.

Reference:
Oral microbiome: Unveiling the fundamentals. *Journal of Oral Maxillofacial Pathology* 2019 Jan-Apr; 23(1): 122–128

Refined sugars and flours, also known as processed carbohydrates, alter our saliva properties creating the conditions that allow erosion of dental enamel and the formation of plaques under the gum line. A nutrition based on whole plants paired with the daily use of flossing protects us against gingivitis (gum inflammation), tartar, and cavities. Since many medical insurance plans do not include dental care and repairs are so costly, the ideal is to keep our original ivories in good condition.

The connection between oral health and heart disease is well established. For example, the mouth microbiota converts the nitrates into nitrites. Throughout the body, the vascular endothelium uses nitrites to make nitric oxide. This gas keeps the vessels elastic, makes blood cells less sticky, and prevents the deposition of subendothelial plaques. Eating plants is an inexpensive habit which helps us build cardiovascular health.

Reference:
Food sources of nitrates and nitrites: the physiologic context for potential health benefit. *The American Journal of Clinical Nutrition* 2009 Jul;90(1):1-10.

THE ESOPHAGUS

The distance between the mouth and the stomach is about 12 inches (25-30 cm). Each bite of food that we swallow travels this stretch in waves of peristalsis, like the ones easily seen in the body of an earthworm. Whether you are an astronaut in space, bedridden, or standing on your head, all solid or liquid we swallow reaches the stomach the same way.

The esophagus ends in a muscular ring (sphincter) that must relax to allow food passage to the stomach. Smoking, drinking alcohol, being sedentary, and eating greasy food or pastries, makes this sphincter relax at the wrong time. Likewise, eating or drinking to the point of filling the stomach or just being overweight, puts too much pressure on this sphincter making it relax. As a result, the acid stomach content travels the wrong way, coating and burning the esophagus.

This problem, which is strongly associated with diet and being overweight, could be relieved with better eating habits, returning to a healthy weight, and practicing some form of daily exercise. Yet, a general public educated by the pharmaceutical industry immediately reaches for drugs to reduce acid production.

According to a global poll, 25% of adults are taking pills to reduce stomach acid for over a year. Did you know that the chronic use of proton pump inhibitors is associated with bone fractures, kidney damage, pneumonia, and some gastrointestinal infections?

Normally, stomach acid kills most of the bacteria we swallow. However, proton pump inhibitors eliminate the production of acid and the symptoms of acid burn, but do not stop the reflux (which now comes with live bacteria). If at night you breathe the stomach reflux that reaches the throat, those live bacteria gain free access to your lungs. This complication is called aspiration pneumonia.

The unnecessary and careless use of pharmaceutical drugs for conditions brought about with food and lifestyle place us at risk of serious complications. We can recover natural protections returning to our native diet of grains, mushrooms, beans, root vegetables, fruits, nuts, and seeds without added oil or sugar.

References:
Dietary Habits and Their Impact on Gastroesophageal Reflux Disease (GERD). *Cureus.* 2024 Jul 27;16(7)

Proton pump inhibitor use: systematic review of global trends and practices. *European Journal of Clinical Pharmacology* 2023 Jul 7;79(9):1159–1172

NEXT STOP: THE STOMACH

When the bolus (a mass of chewed food) passes to the stomach, it drops into a super acid chamber where most bacteria die and food continues its decomposition. As the links between amino acids dissolve, proteins unfold and enzymes break them into smaller and smaller pieces. If the protective lining or mucosa is too thin or worn, the acid will come in contact with the naked stomach wall and will want to digest the gastric cells, causing the characteristic ulcer pain or burning.

The same conditions that increase the risk of esophageal reflux increase the risk of gastritis. This chronic inflammation correlates with consuming fried food, animal products, being overweight, use of tobacco, lack of restorative sleep, stress—in other words, with the western lifestyle. The use of anti-inflammatory drugs, such as ibuprofen, naproxen, aspirin, and steroids increase the risk even more.

THE DUODENUM

After 20 to 90 minutes, food crosses another muscular ring at the end of the stomach, upon which the pancreas immediately discharges bicarbonate to neutralize the stomach acid. Without this automatic protective mechanism, the acid produced by gastric cells would eat up duodenal cells causing an ulcer with a penetrating pain that gets localized behind the navel or in the back.

Foods rich in antioxidants, vitamins, and phytochemicals protect the integrity of the intestinal tract and help keep the functional competence of the muscular rings at the entrance and exit of the stomach (esophagogastric and gastroduodenal sphincters).

If you ever had stomach pain, did you wonder why or automatically reach for a pill?

We have a muscle that separates the thorax (chest) from the abdomen (belly), it is called the diaphragm. Consider it the chest's floor and the abdomen's ceiling, so to speak. Lungs and heart occupy the superior cavity, while stomach intestines, and other organs occupy the bottom one.

The diaphragm contracts and relaxes with respiration. Its movement is like the bellows of an accordion that help fill and empty the lungs of air. When the diaphragm is relaxed it has the configuration of a dome, compresses the lungs and helps them blow out air. When it contracts during inspiration, it turns into a flat shelf that presses against the belly, which makes the chest cavity larger, helping the lungs fill up with air. Watch yourself in a mirror without tightening any belly muscles and see how your torso moves with each breath.

Being overweight especially around the waist, eating until full, or during late pregnancy, all tend to compress the stomach putting pressure on the gastroesophageal sphincter. Also drinking alcohol, refined flour confections, and fritters by themselves relax the sphincter and increase the risk of reflux. The most important thing you could do to improve this condition is adopt good plant eating habits.

Much less commonly, the symptoms of reflux stem from a paraesophagic hiatal hernia. At the center of the diaphragm is a small hole or hiatus through which the esophagus connects to the stomach. That hiatus functions like a snug buttonhole. It should not be too large, otherwise as you bend over to pick up something the pressure might push part of the stomach through it up onto the thorax. This abnormal configuration prevents the sphincter from fully closing and reflux ensues. Depending on the severity of the symptoms, a paraesophagic hiatal hernia might require surgical repair.

THE PANCREAS

The pancreas is shaped like a long fish, weighs as much as a deck of cards, and is located behind the stomach. It is famous because it produces insulin: the hormone that facilitates glucose entry to the cells where the mitochondria convert it to energy. Every mammal, reptile, and bird has a pancreas that produces the insulin hormone with the same purpose: to facilitate the transport of glucose across the cell membrane, so the mitochondria inside the cells can convert it into energy. It is impossible to fully replace the sophisticated interactions that exist between the insulin, the pancreas, the glucose, the intestines, the brain, and the cells of the body. The management of type 1 diabetes, where the pancreas produces no insulin and the person must inject it every day, presents great challenges. However, with the discovery that insulin resistance operates in all types of diabetes we gained great insight and a very useful tool.

One of the most effective strategies for the successful treatment of diabetes type 1 and for the possible reversion of type 2 is to focus on reducing insulin resistance. The resistance to insulin is independent of whether the insulin is endogenous (produced by one's own pancreas) or injected (as in the case of type 1 diabetes).

Insulin resistance can be reduced following a dietary pattern based on plants that contributes less than 30 grams of fat daily. We will talk about diabetes later on.

> References:
> A plant-based diet for the prevention and treatment of type 2 diabetes. *Journal of Geriatric Cardiology* 2017 May;14(5):342–354
>
> *Mastering Diabetes: The Revolutionary Method to Reverse Insulin Resistance Permanently* Cyrus Khambatta PhD and Robbie Barbaro MPH

Besides producing insulin, did you know that the pancreas discharges digestive enzymes directly to the intestines?

Enzymes are special proteins that facilitate biochemical reactions. They can act in different processes and can be reused without being consumed. The pancreatic enzymes help complete the digestion of the fats, proteins, and starches in our foods.

Lipases. These enzymes act on fats (grease) and work together with the bile produced in the liver to facilitate the absorption of vitamins that must be dissolved in fat to be incorporated into the blood (A,D,E,K). When fats are not absorbed, they are eliminated in greasy diarrhea.

Proteases. These enzymes help to decompose proteins in the diet. They also destroy the proteinic cell wall of bacteria, protecting us from disease-causing germs that we inevitably swallow and might end in the intestines. Animal proteins that are not fully digested are absorbed into the blood as recognizable fractions that act as antigens or foreign bodies. These antigens stimulate the immune system to make antibodies. These antibodies are later capable of attacking our own proteins in what are known as autoimmune diseases.

Amylases. These enzymes help break starches that will be turned to glucose. When we lack amylase to digest these carbohydrate starches, we produce watery stools.

As food travels down the intestine, more and more chemical bonds continue to break until the bolus turns into a semiliquid amalgam. The fatty portion passes to the lymphatic system, which takes it through nodes or stations that filter it free of bacteria and any other harmful component, before allowing it to enter the circulation. When the digested units are sufficiently small and capable of going through the intestinal barrier, they get absorbed into the blood.

The bolus that continues down the intestine is obviously dragging undigested material. Some are toxins that we want to discard, drug metabolites, cholesterol that was successfully removed from the arteries and the liver emptied in the bile, plus dangerous chemicals, and much more.

Traveling together with all this waste we find a complement of carbohydrates (cellulose, pectin, hemicellulose, and lignin) that resist our digestion. We call these resistant carbohydrates, fiber. Most animals lack the necessary enzymes to digest these valuable carbohydrates, so they reach the colon intact. There is where they meet the microbiota or intestinal flora—about five pounds of bacteria, protozoa, fungi, and yeasts—that do possess the necessary enzymes to digest fiber. In fact, this population has settled in the intestine precisely because our plant rich diet provides them a constant source of food. Their successful digestion or fermentation of these resistant carbohydrates supplies us with highly nourishing short chain fatty acids which we will discuss later.

THE LIVER

The liver is an organ that weighs around three pounds; it is located on the right side of the abdomen protected by the rib cage. Among its many functions, it filters the blood of cholesterol, alcohol, some drugs, and a great many toxins. Together with fat emulsifiers, the liver discharges these waste products onto the gall bladder, where they are temporarily kept until they are squirted to the intestines and carried down to be excreted. High concentrations of cholesterol in the bile favor the formation of gall stones.

The liver is subject to damage by alcohol, environmental organic solvents, and some drugs such as acetaminophen. It also stores vitamins, such as A, B_{12}, D, E, K, and minerals like iron and copper. The liver is very important in the metabolism of glucose and works best when its cells are not filled with fat.

When I was a student, fatty liver was a common condition among people who drank alcohol. Since my medical school years, the western diet has become the number one cause of fatty liver. Chronic liver inflammation causes fibrosis, scarring, and hepatic failure, which carries 25% to 30% mortality. Regardless of whether the cause is alcohol, a viral infection, or diet, 90% of liver insufficiencies can be avoided.

THE MICROBIOTA

The microorganisms that reside in the colon are descendants of those who luckily escaped alive when they passed through the stomach. These microbes traveled the lethal hydrochloric acid lagoon sheltered in the stem of Swiss chard, a cauliflower floret, the wrinkles of carrots, the folds of kale, pineapple seeds, the skin of grapes, the soil stuck outside a beet root. From their hiding places, these live entities or probiotics saw horrified how their companions died dissolved in the stomach acid.

After the trauma, they continued their uneventful travel through the small intestine (duodenum, jejunum, and ileum) until they reached the large intestine (colon), where they joined some five pounds of commensals called the intestinal flora or microbiota. If we feed them, they will multiply and prosper. Otherwise, they die.

The microbes in the colon ferment the carbohydrates that resist our digestion (fibers) and produce propionate, acetate, and butyrate. These short chain fatty acids contribute between 5% to 15% of our daily caloric requirements and between 60% and 70% of the energy that colon epithelial cells require. It is not just worth our while to give these organisms shelter, since they pay good rent! I also want to reward them and help them live.

In general, we could say that there are two populations of microbes or probiotics in the colon. Some specialize in fermenting carbohydrates that resist our digestion and produce beneficial short chain fatty acids that fertilize the colon cells, feed our brains, and improve the function of our immune system. Others prefer to digest animal protein and fat, eggs, cheese, and fritters generating toxic molecules (trimethylamine) which cause inflammation throughout the body and which the liver oxidize and turn into trimethylamine oxide (TMAO), a molecule associated with heart attacks.

This population of more than 100 trillion unicellular organisms is in a constant flux. They grow depending on what we eat. Which ones win? The ones we feed. Do you remember reading that our bodies are made of 37 trillion cells? This translates to the fact that the number of microorganisms in our colon is even larger than the number of cells in our structural body.

Prebiotics feed the probiotics who in turn produce postbiotics (you identify them by the prefixes pre, pro, and post).

We acquire the first probiotics that ever inhabit our intestine when we pass through the vaginal canal. This seeding is so important that today babies born by cesarean section are given an inoculum of bacteria from the mother's perineum.

Food comes with its sample of probiotics. So, our diet continually adds new microbes to our intestinal flora. The probiotics that eat the fiber of a purple sweet potato come on the skin of purple sweet potatoes, the ones that ferment black bean fibers come in a spoonful of black beans, and so on. We simply must feed on a variety of plants every day.

Only one in ten people have a real food allergy, however around one in five believes to be allergic to some food. More commonly, the problem is a deficiency in the macrobiotic bacteria that ferment the food in question, and the person notices that a certain variety of foods triggers similar symptoms. The point is that if we have decimated a population of bacteria, by whichever means as we'll see below, it could be months before they can be restored to a viable bunch if not lost forever, but do not despair.

Here is the importance of regularly consuming the greatest variety of food. According to many gastroenterologists, we should strive to eat 30 different plants every week. You easily approach this amount when you include different types of mushrooms, pumpkins, rice and beans, different varieties of sweet potatoes, onions, radishes, cabbages, and beets. The greater variety of plants, the more beneficial bacteria are introduced to the intestine. If they are not readily accepted at first, reduce the quantity, but do not eliminate them!! Act with the certainty that you are feeding an army for the general benefit of your immune system.

Commercial supplements that are labeled as probiotics only provide a small sample of the immense variety of microbes present in whole plants and they do not come with any food for them. Remember that if you do not feed them, you have wasted your money.

The indiscriminate use of antibiotics kills beneficial microbes that may never have a chance to recover, but also allows pathogenic bacteria resistant to such antibiotics to survive and multiply without competition. Should you need to take an occasional antibiotic, try to restore your microbiota the way described above. If for some reason you require frequent antibiotics, the picture is more complicated. In such cases, commercial prebiotics and probiotics might be helpful and you should probably seek help. It will not be the same as eating whole plants, but do not throw the towel.

THE INTESTINAL LINING AND MUCOSA

Before crossing the intestinal barrier, food spends hours preparing itself to be worthy of becoming part of the body (being incorporated). For someone whose diet is plant rich, the intestinal transit is completed in 7 to 14 hours. In contrast, intestinal content lacking fiber can take up to 72 hours, allowing discarded cholesterol and other metabolites and toxins that the body had discharged onto the bile to be excreted, to be absorbed right back into the blood. What a waste of valuable work!

> Reference:
> HDL cholesterol efflux capacity is inversely associated with subclinical cardiovascular risk markers in young adults: The cardiovascular risk in Young Finns study. *Scientific Reports* volume 10, Article number: 19223 (2020)

The inside surface of the intestine has the appearance of a fine towel or velvet, where the sum of each fingerlike projection or villus increases the functional intestinal area to 4,000 square feet (370 meters square) Along the digestive system, its surface is coated by two mucus layers, one adhered and one floating. These layers protect the intestine from toxins and other harmful substances, while they also facilitate the passage of nutrients to the blood. The bacteria in the colon (the small intestine is almost devoid of bacteria) live within the floating mucus layer where they find shelter and a grip to avoid being dragged with the bolus of food that keeps moving toward the exit.

The intestinal membrane must maintain a tight union between its cells, this keeps substances that could negatively alter the immune system from entering the blood. Think just how amazing this is: since food traveling the intestinal canal is technically still outside of the body, there is only one layer of cells separating what we put in our mouths from the blood that circulates our system. Think of how we abuse and stress this marvel of ours.

Reference:
Gut microbiota, intestinal permeability, and systemic inflammation: a narrative review. *Internal and Emergency Medicine* (2024) 19:275–293

While studies and discoveries continue to flesh out the details, the association between the western American diet and elevated inflammatory markers is unquestionable. The western American diet is rich in fats, animal protein, and refined carbohydrates such as flour and sugar. It results in elevated inflammatory markers such as C-reactive protein, tumor necrosis factor, interleukin 6, and proinflammatory cytokines.

Chronic inflammation in someone who eats meats, cheese, eggs, fritters, pastries, food high in sugars and oils produces a lot of waste in the form of free radicals. This weakens the unions between the cells of the intestinal membrane and creates a permeable or leaky gut that can allow antigenic proteins to seep into the blood. The immune system uses these antigenic particles as molds to produce antibodies.

One of the many advantages of consuming proteins of plant origin is that since they don't look at all like our own, any antibodies made to their image cannot be used against our proteins. Whereas, antibodies made against animal protein or animal protein fragments (we are also animals) can unleash an autoimmune response . We witness this in autoimmune diseases like rheumatoid arthritis, type 1 diabetes, multiple sclerosis, lupus, thyroiditis, and to a certain extent Crohn's disease.

Reference:
Plant-Based Dietary Patterns for Human and Planetary Health. *Nutrients* 2022 Apr 13;14(8):1614.

Regarding inflammation, a nutrition based on plants not only reduces free radicals formation, but its antioxidant properties neutralize any free radical excess. By reducing inflammation, we also help maintain a healthy union between the cells in the intestine and avoid a leaky gut.

These discoveries give us hope, because they indicate that we can modulate even the autoimmune diseases that are so devastating. By practicing plant-based nutrition when we are still healthy, we can avoid much pain and suffering. A stitch in time saves nine.

It seems crazy to speak of the body as if it was a machine made up of parts, but I find no other way to highlight some aspects of our organism without temporarily isolating them from the whole. Mind you, after studying them like so, we must humbly put it all in context and marvel at how we function as a whole. Remember that the whole is always more than the sum of its parts. We will do the same with food when we get to the section of nutrients.

THE KIDNEYS

These two magnificent organs filter the blood we have in circulation sixty times each day. Two hundred liters of blood pass daily through the kidneys who in turn produce two liters of urine (enough to fill a small swimming pool in one year). Urine is mainly water in which 2,600 species of metabolites have been dissolved. Among them are inorganic salts, urea, uric acid, and many other waste products. Urine is filtered from sterile blood, therefore urine is also without bacteria.

The urine metabolome is a subgroup of the blood; in other words, anything we find in the urine must also be present in the blood. Take for example nitrogen (N) which is a characteristic element of protein molecules; when there is excess protein in the diet, we eliminate N in the urine.

Having said this, it was surprising to identify 484 compounds that had not been previously reported in blood. This could only mean that the kidneys not only filter blood, but they extract undesirable metabolites in concentrations below what we can detect in blood and concentrate them in urine up to 1,000 times.

The kidneys discriminate between what they want to keep what they want to discard. They retain valuable substances, such as calcium and proteins, and let go of ammonium and other poisons. Many waste products are eliminated via the liver, but others such as antibiotics, insulin, and some analgesics are eliminated by the kidneys. If they lose their ability to function, some drugs can accumulate to toxic or even lethal levels.

As we enjoy a diet based on whole plants, we also reduce the amount of drugs that go into our system and lighten the burden for these marvelous organs.

Reference:
The Human Urine Metabolome. *PLOS Public Library of Science*
Published: September 4, 2013

THE UTERUS

The uterus is a hollow muscular organ inside which a fetus could develop. Other large muscular organs with a cavity are the urinary bladder and the heart. According to world statistics, 70% of reproductive age women have benign tumors or myomas in the uterine wall. These tumors are associated with estrogen, chronic inflammation, and obesity. They are more common in women who consume meat and ham and rare in women whose diet is rich in vegetables, cruciferous, soy, and fruit. Given that the only permanent treatment that medicine has to offer is surgical, following an anti-inflammatory diet based on plants seems like a benign and logical non-invasive option.

THE ARCHITECTURE OF THE SKELETON

The frame of our body consists of 206 articulated bones. The ends of each bone are covered with a specialized glossy white articular cartilage. This connective tissue protein provides a smooth lubricated surface that allows one bone to move against another and transmit tension forces with the least amount of friction. The articular cartilage has a very limited capacity to repair itself since it has no vasculature, and receives oxygen and nutrients from the surrounding synovial liquid by diffusion.

A nutrient rich diet assures a nourishing synovial liquid.

We should not be surprised that the knee is the joint that has suffered most with the obesity epidemic. Besides representing a suboptimal diet, the excess weight compresses the joint cartilage, squeezing the water out of it and making it more brittle.

This inner scaffolding of bones is held together by tendons and ligaments. Our skeleton is like a puppet that moves by cords and pulleys made of collagen tied to over 600 muscles made of protein fibers.

Our organic architecture and that of all animals is made of collagen. This protein corresponds to one third of the total protein in the human body and is the most abundant component of our extracellular structure or matrix. When we speak of the body, we are referring to this whole corporeal structure made with proteins.

Although they are flexible, ligaments barely stretch and tendons should not stretch at all! Collagen is resistant to tension and compression. The muscles stretch and contract allowing for flexibility across joints. The inorganic components of our bones are rigid.

> Reference:
> *The Nutrition Business Journal 2022* reported a robust growth in the collagen market that has already reached $222 million and maintains a focus on any and all concerns shown by the public—hair, skin, nails, joints, and general wellbeing.

There is no investigation on collagen supplements to date that is independent from industry. Where will you find credible non propaganda information on which to base your decisions?

The truth is that a diet of enough grains, beans, fruits, vegetables, mushrooms, nuts and seeds supplies all the amino acids for the body to make its own collagen. Commercial collagen supplements are made of exactly the same amino acids that are found in plants.

We can protect the collagen of our bodies by eliminating tobacco and avoiding prolonged sun exposure—risk factors that contribute to collagen destruction.

MUSCLE FIBERS AND HOW THEY GROW

Muscles are made of many different types of proteins. We can increase the diameter of any muscle with six weeks of resistance training. When a muscle hypertrophies, each myofibril (muscle fiber) divides lengthwise, making for more fibers within each muscle fiber. Imagine the myofibrils as the threads or filaments that make up a rope.

References:
Therapeutic Exercise / Muscle Hypertrophy. *Braddom's Physical Medicine and Rehabilitation (Sixth Edition) 2021*

Uncomplicated Resistance Training and Health-Related Outcomes: Evidence for a Public Health Mandate. *Current Sports Medicine Reports* 2010 Jul-Aug; 9(4): 208-213

Resistance training stimulates muscle growth much more than aerobic exercise. Besides helping to slow muscle loss in the aging person, it reduces physical disabilities, falls, and even the risk of premature death in general. This type of exercise done daily strengthens and keeps muscles elastic which also improves balance.

We have correctly promulgated that resistance training develops muscles and muscle strength, but we have incorrectly stated that it does not promote cardiometabolic health. In fact, it has been demonstrated that all types of training (resistance and aerobic) are cardio protective. It should be clear that all types of exercise contribute to the country's better physical and mental health.

Reference:
The 4 most important types of exercise: strengthening, stretching, balance, and aerobic. *Harvard Health Online*

Using the wrong information affects your decisions and their outcomes. Perhaps you have been told that riding a bicycle or swimming do not contribute to bone health because it is not weight bearing or against gravity. Signing on to this myopic notion, you park the bicycle in the garage, put away your swimsuit, spend money on spandex and a trampoline, and off you go jogging and jumping. It is true that such exercises stress the bones into retaining calcium. But what happened to the thousand and one benefits of the other exercises? How many did you throw away?

The lens of reductionism is not a practical one. These pages present what we know today, not the last word. When we think we have arrived, a gap opens and we drop to another level where our eyes see a fresh new place that demands we investigate some more. Don't ever stop being curious.

THE BRAIN

The brain is a gelatinous mass of three pounds, composed of 86 billion neurons and one trillion support or glial cells with over a trillion connections. Although the brain lives protected in a vault, the nervous system extends its tentacles, 45 miles of nerves, that send electric impulses messaging throughout the whole body.

To maintain the flow of electrons channeled and swift, each nerve fiber is insulated with a myelin sheath, made of lipids and proteins. The integrity of this capsule is assured by a whole plant diet with adequate levels of vitamin B_{12}. In multiple sclerosis, myelin disintegrates due to autoimmune inflammatory damage.

The millions of neuronal connections in our brain are formed throughout our lives with the help of reliable sources of nutrients and energy and according to one's learnings and experiences. Our responsibility toward not just the brain but every organ is to provide them what they need without putting obstacles in their way. When we study the so-called Blue Zones of the planet, we see that the best way to fulfill that mission is with a whole plant-based nutrition low in fat, exercising the body regularly, sleeping around seven to nine hours, without tobacco or other harmful substances, free of toxic stress, and living with purpose in community.

SLEEP AND DREAMS

Every night we fall unconscious for seven to nine hours, unable to move voluntarily. Truth be told, we all must surrender to sleep, it's written in our DNA. It seems that we must alternate between being awake and being asleep to function optimally and to capacity. There is evidence that during the hours of sleep we detoxify the brain of inflammatory and oxidative waste that has accumulated during the day.

Something curious also happens during those unconscious intervals of what we call REM (rapid eye movement). Most of us remember having intense emotions and vivid sensations: scenes as from the theatre, live images, conversations, places, colors, smells, sensory experiences. I have journals filled with notes from these expeditions. The first dream that I remember was when I was twelve years old; I was in a wagon going over dry dusty soil. Many dreams are fantastic voyages, others tell everyday ocurrences. Almost every night they delight me, surprise me, and guide me. The characters, mostly unknowns, don't scare me anymore. They are mine, they come from my interior, my unconscious. It seems that we can develop an ever-growing capacity to remember more details. If as soon as we wake, we jot down any shred even a thread of what we dreamt, we cultivate the practice of remembering a journey that carries us to a different adventure every night.

It is also thought that sleeping and dreaming help us in part to organize thoughts and memories. Recently, Eric Hoel, a neuroscientist from Tufts, has proposed that perhaps the greatest use is in helping us learn. He theorizes that dreams serve to improve the function of generalizing and to avoid overfitting. Overfitting refers to the ability to foresee and respond with certainty in situations for which we have trained, but is inadequate when presented with new data or information.
Hoel adds that we also develop this function when we expose ourselves to works of fiction, novels, movies, art hung on the walls of a museum, or murals on city buildings. We need to find an equilibrium between the tendency to feel content that we know a route with eyes closed (overfitting) and the valuable ability to come out of our comfort zone and aptly journey through unknown places (generalize). Perhaps that is why we enjoy so much those detours that take us away from the routine.

It appears that the more connections we make between neurons, the more connections we can spare over the course of our lives without the loss of cognitive function typical of Alzheimer's (a progressive neurodegenerative condition characterized by memory loss, thought disturbances, and behavior changes).

We need dreams, but we do not need to understand them. If it was necessary to interpret dreams for them to be useful, how do we explain the fact that all mammals—dog cubs, young elephants, and human babies—all dream? Why don't reptiles dream? According to the theory, mammals need to learn, whereas for the iguana and the turtle it is all innate. As soon as she is out of her eggshell, the turtle makes a beeline to the sea and swims. We say they act on instinct, they don't need anyone to show them the way. When I was a little girl, digging in the soil, I found countless lizard eggs—half inch soft sacs and white as snow—from where I saw little tiny lizards come out running after a moth, perfectly formed, already enroute and with a mission.

> Reference:
> A New Theory for Why We Dream; Investigator from Tufts, Erik Hoel. *Neuroscience* February 18, 2021

I want to protect my sleep because it is something important needed by my body and my brain; I don't want to consider it an inconvenience that interrupts my "productive" hours. I want to establish a routine where I voluntarily make room for sleep, where I honor this necessity instead of debasing it or dominate it with caffeine. I don't need to totally understand the function of sleep to realize that, just as food, water, and air, I cannot do without it.

Everything that benefits or hinders the body is also operating on our psyche. It is quite reductionistic to speak of mental health as separate from the rest of the body. Our imagination can help heal that artificial gap we have created between body and mind.

WHO AM I?

Spread out, our envelope is 2 two square meters (21 ft^2), 10 kg (22 pounds), and sports several holes.

The skin is flexible and stable at the same time. As a landscape it is dry, acidic (pH between 4.0 and 6.0) and without many nutrients, perfect to rebuff any pathogen that might attempt to colonize it. The skin opens the body to the outside through six or seven well-guarded apertures: eyes, ears, nose, mouth, urethra, vagina, and anus. Each of those holes is populated by microbes specialized for the habitat location. We coexist with more microorganisms than the number of cells that make up our bodies. We care for the skin the same way we care for the rest of our selves: with excellent nutrition, not smoking, and protecting it from excess exposure to sun which prematurely ruins its collagen.

THE LUNGS

We live within Earth's atmosphere, surrounded by air from where we extract oxygen and to which we discard carbon dioxide. This gas exchange takes place in the most hidden places of our lungs. The air which enters the trachea through our nose and mouth continues through two bronchia that themselves divide like the branches of a tree, until they end up in alveoli (little vascularized sacs within the pulmonary tissue), where air and blood are separated by only one cell.

The respiratory tract requires conditions of optimal temperature and humidity to operate. Depending on the air quality, we expose ourselves to more or less contaminants.

The respiratory tract is lined by a thin mucus and a sticky phlegm to which waste particles that enter with respiration can adhere. The mucosa or inner respiratory lining is covered by villi that move like a fan, steadily like a wave, in one direction toward the outside. This keeps us from drowning in mucus, phlegm, and contamination. Smoke is hot, it burns and shortens these hairs incapacitating their function. The risks of cigarettes are not limited to cancer—the published scientific studies have finally reached the public.

THE LYMPHATIC SYSTEM

Unlike the circulating blood, lymph travels in one direction. The lymphatic system transports between eight and twelve liters of lymph; produces white T and B cells; recovers liquids from the interstitial spaces (between tissues); facilitates the absorption of fats from the digestive system, and optimizes the immune system through the mobilization of macrophages, lymphocytes, and other cells. This system includes the thymus, spleen, tonsils, and approximately 450 lymphatic nodes. Within these stations or filters, lymphocytes interact with bacteria, viruses, and other particles that may have entered our body and fabricate antibodies. Once it has passed through one of these nodules, the clean lymph enters the blood through a lymphatic duct.

The recovery of lost plasma: Life subjects us to traumas which break capillaries under the skin (small superficial vessels). As a result, there is a spill of blood to adjacent tissues and a purple or black blotch stains the skin. For the better part of a week this discoloration changes from black and blue to yellow green due to the disintegration or transformation of hemoglobin into bilirubin. In time, the components are absorbed by the lymphatic system and the blotch disappears.

Reference:
Bruise dating using deep learning. *Journal of Forensic Sciences* 2021 Jan; 66(1): 336–346

THE CARDIOVASCULAR SYSTEM

Different from the unidirectional lymphatic system, the one which transports blood is a circulatory system, in other words, it circulates coming back around.

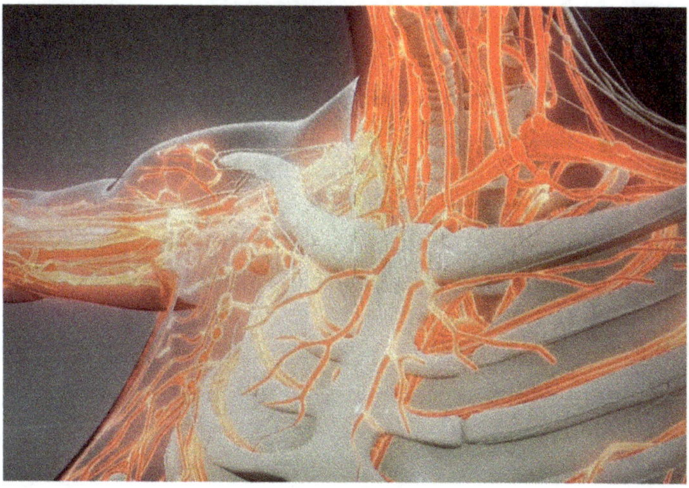

This system consists of miles of blood vessels that we must keep clean and elastic.

The cardiovascular system is a closed circuit with five liters (slightly more than a gallon) of circulating blood that travels a highway which if placed end to end measures 64 thousand miles long (103 km). The arteries take blood out of the heart and the veins return blood to the heart. Each one of our cells receives oxygen and nutrients thanks to this delivery service.

Have you ever wondered how blood vessels get nourished? Vasa vasorum or the vessels vessels originate inside the artery or vein and penetrate the vascular wall to feed it with nutrients and oxygen. The external vasa vasorum originate from a nearby vessel branch and supply nourishment to the larger vessel wall. Capillaries, on the other hand, are so thin that they are nourished by diffusion. Every cell, without exception, needs food and oxygen.

The blood hauls nutrients we derive from digestion, such as glucose, amino acids, triglycerides and other fats, phytochemicals, vitamins, and minerals besides all that we fabricate with them like hormones, red blood cells, white blood cells, and platelets. The blood also transports medications and waste together with oxygen and carbon dioxide in solution. All of this circulates in a sterile aqueous medium down 64,000 miles of blood vessels (arteries, veins, and capillaries).

The heart has two exit doors: blood is taken through the pulmonary arteries to the lungs to get oxygen and the aorta sends oxygenated blood out to the brain and the rest of the body. For the tissues to exchange gasses, nutrients, and waste the arteries need to branch until they are as thin as a thread (capillaries) where blood cells travel single file. These highways must stay clean and elastic. Keep in mind that when cholesterol plaques are forming in the arteries that supply the heart (coronaries), they are also forming in arteries throughout the body.

The inside of blood vessels is lined with a very thin layer of cells called the endothelium which produces nitric oxide, a gas that among its many other functions keeps the vessels elastic. An endothelium damaged by too much alcohol, salt, saturated fat, or cholesterol plaques cannot produce enough nitric oxide. Arteries that cannot dilate to deliver oxygen to a muscle during exercise cause angina in the heart or pain in the chest, angina in the calves when walking uphill, abdominal or mesenteric angina after eating, back pain due to oxygen insufficiency in the vertebral column, or sexual erectile dysfunction.

During sexual excitation, nitric oxide relaxes the artery that supplies blood to the penis. The dilated artery compresses the vein in the shaft causing something like a tourniquet. The blood temporarily stuck in the penis turns it hard with an erection. Did you know that the famous pills for sexual dysfunction (a modern euphemism for impotence) are nitric oxide donors that dilate the penile artery?

The nitroglycerin that some people use under the tongue during an angina episode is also a nitric oxide donor. How are you going to address pain or dysfunction? With a pill? With bypass surgery? The body sends you valuable messages. Wouldn't you prefer to go to the cause? Wouldn't it be more appropriate to assume a change in diet that could reestablish elasticity of the arteries and reduce the cholesterol plaques? Wouldn't it be wiser to help heal the endothelium. That way you could produce more nitric oxide.

I could bet that your health team does not even mention these options. This is because most physicians do not know enough nutrition or the power that plant-based nutrition has to revert coronary artery disease and also because interventional medicine is too lucrative. My dear colleagues, how will we correct this situation?

References:
Peripheral Arterial Disease Masquerading as Low Back Pain. *American Journal of Physical Medicine & Rehabilitation* 91(12):p 1104-1105, Dec 2012

Assessment and management of peripheral arterial disease: what every cardiologist should know. *British Medical Journal Heart* Volume 107, Issue 22

Mesenteric Arterial Insufficiency and Abdominal Angina. *Archives of Internal Medicine* 1964;114(6):765-772

Real-World Clinical Outcomes of Indigenous Biodegradable Polymer Drug-Eluting Stents. *Cureus.* 2021 Sep; 13(9): e17886.

There are emergencies that warrant placing an endovascular prosthesis or stent to reestablish blood flow and oxygen to a compromised area. However, after restoring the circulation, the same stent material stays in the artery causing an inflammatory reaction that could delay healing of the endothelium, increasing the risk for thrombosis and restenosis. Stents of biodegradable material and which decompose in a few years, will be a viable option in the near future. Having said this, we don't need to wait. A logic alternative is to adopt a diet and lifestyle that delays the progression of atherosclerosis in native arteries as well as in a stent. With the same diet and lifestyle, we could aspire to even reverse established disease.

Reference:
Prevent and Reverse Heart Disease: The Revolutionary, Scientifically Proven, Nutrition-Based Cure. Caldwell Esselstyn, MD

Myocardial infarctions or heart attacks are the number one cause of death in countries that follow a western dietary pattern. They occur without warning in arteries that are beginning to develop plaques. In other words, in arteries with young plaques that have not occluded the lumen enough to be symptomatic. According to published studies,

90% of infarcts can be prevented. However, from the moment you begin sending clean blood through the circulatory system, the disease begins to slow down and even reverse. You don't need permission from a specialist to stop smoking or to adopt a whole plant diet. You don't need approval or medical insurance to clean your kitchen of meat, milk, cheese, eggs, oil, and junk food. If you require a shopping guide, recipes, or support, go get it. Don't allow these obstacles to get between you and your health.

The Center for Nutrition Studies T Colin Campbell, the Physicians Committee for Responsible Medicine, (PCRM) the American College of Lifestyle Medicine (ACLM) and other not for profit plant-based organizations offer webpages with many free recipes that could help you.

ADIPOCYTES

Among the 37 trillion cells of the human body, there are generalists and specialists. The adipocytes are a group of specialized cells of great interest. We primarily know these cells because they manage our fat stores, cushion and shield our internal organs from injury, and give us our characteristic curvilinear silhouette. By specializing in the storage of fat, they protect muscle and liver cells, which prefer to work lean (free of excess fat).

The adipocytes play an important role in non-communicable diseases, because when these cells overfill with fat they expand to the point of a split or crack spilling its fatty acid content to the surrounding tissues. When oxidized triglycerides leach out of the adipocytes, they are toxic. The accumulation of tissue damaging molecules known as free radicals unleashes the chronic inflammatory chain reactions that are at the core of diseases such as atherosclerosis, diabetes, arthritis, Alzheimer's, and cancer.

In other words, obesity is not benign, instead it is a trigger for the most common and feared diseases that wear and tear our health.

PART TWO

Food and its Nutrients

Through the process of photosynthesis, plants take CO_2 from air, minerals from the soil, water from rain, and light from the sun to make the carbohydrates, proteins, and fats that power the planet and in turn release vital oxygen to the atmosphere. Plants make their own tissues from scratch. Whereas, we animals must eat.

CARBOHYDRATES

Second to water, carbohydrates are the most abundant nutrient on the planet. A third of the animals on Earth evolved with enzymes that digest the carbohydrates found in roots, stems, leaves, mushrooms, fruits, and seeds. These digestive enzymes break the bonds that tie together the nutrients in carrots, spinach, rice, corn, wheat, oats, quinoa, lentils, garbanzos, beans, peanuts, cabbage, cauliflower, mango apple, grapes, tamarind, papaya, almonds, walnuts, pistachios, sesame seeds, flaxseeds, chia, and other food that grows out of the ground.

Among all carbohydrates, glucose is the simplest one: a monosaccharide whose formula is $C_6H_{12}O_6$. Glucose enters the body through plants in the form of other carbohydrates: galactose and fructose (monosaccharides), lactose and sucrose (disaccharides), or starch (polysaccharides). Glucose is the energy source for all cells and organisms on the planet, it is our fuel.

Normally the human body obtains around 40% to 60% of its energy from glucose. Through many other reactions, we also derive energy from fats and amino acids. The digestible carbohydrates provide glucose for immediate use while we store a certain amount in the form of glycogen (a polymer of glucose) for later use. In the fasting state, we convert stored glycogen back to glucose.

Animals like the polar bear, tiger, wolf, and cat hardly eat plants. They use the protein and fat in the body of other animals to operate their metabolism. They also can undo animal protein, discard the nitrogen as urea, and use its amino acids to synthesize glucose. This process is called gluconeogenesis.

Considering how vital glucose is, it should not surprise us that we can obtain it in various ways.

FIBER

In addition to the digestible carbohydrates we use for energy, plants make other carbohydrates that most mammals cannot digest, since we lack the enzymes to break them up. These carbohydrates are known as fibers (celluloses, hemicelluloses, lignans, oligosaccharides, pectines, gums, waxes, and certain starches). Some of these compounds impart rigidity to the plant structure, while others dissolve in water and form a sort of amalgam.

Having said this, in our large intestine we have a population of microorganisms whose enzymes can break this type of resistant carbohydrate. Their digestion or fermentation of fiber yields short chain fatty acids—acetic, propionic, butyric, and others—for even more nutrition and energy.

> Reference:
> Dietary fiber: moving beyond the "soluble/insoluble" classification for monogastric nutrition, with an emphasis on humans and pigs. *Journal of Animal Science and Biotechnology* 2019;10: 45

People who eat little meat and more plants have a lower risk of colon cancer. It was speculated that the protection was due to the fiber accelerating the intestinal transit and dragging more toxins. Today we have more pieces in this puzzle. It so happens that the short chain fatty acids we have mentioned not only nourish the brain and strengthen the immune system, but also directly nourish the colon cells and protect them against cancer. This sophisticated model of collaboration based on carbohydrates suggests that plants are, and have always been, the main source of nutrition of our species.

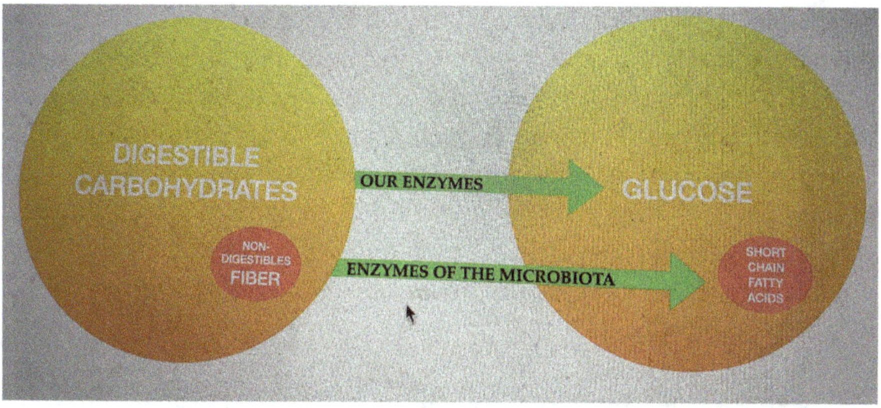

We convert digestible carbohydrates into glucose; the carbohydrates we cannot digest and that our microbiota converts into short chain fatty acids we call those, fiber.

The carbohydrates are one of the macronutrients. (The other two macronutrients are proteins and fats.) This group of nutrients includes molecules that we can digest, such as galactose, fructose, lactose, and sucrose plus others non digestible ones that we call fiber, such as cellulose, hemicellulose, lignines, oligosaccharides, pectines, gums, waxes, and certain starches.

We convert digestible molecules into a more simple carbohydrate called glucose and we use it for fuel. The microbiota ferments the non-digestible ones into short chain fatty acids, such as acetic acid, propionic acid, and butyric acid.

PROTEIN

Protein molecules are chains of amino acids, which alongside lipids form the structure of our bodies. Proteins are an integral component of all organs, muscles, and humors. The 37 trillion cells of the body, the thousands of enzymes, and countless hormones are made of proteins. Proteins are not stored, they are an integral part of our structure and function.

During protein digestion (no matter whether from plants or animals), the bonds that tie the amino acids together break down, liberating each little building block or amino acid to be incorporated in the synthesis of the next human protein. If you consume enough carbohydrates for your energy needs, you will not need to digest your muscles made of protein. People who do not consume enough calories from carbohydrates have had to burn their body, invertedly losing weight from muscle mass. Think of the emaciated people living in war zones, in extreme poverty, persons with a terminal illness, or those with an eating disorder like anorexia. The key to getting enough nutrition is in consuming enough food—the nutrients come with it.

Plants make their proteins using just twenty amino acids, the same twenty that we use to make the proteins of the human body.

We create the hundreds and thousands of proteins required by the body using twenty amino acids. We build all the words in the English and in the Spanish vocabulary with only 26 letters; and we compose an infinite number of melodies with an octave of 12 notes and half notes—it's no different.

We have 20,000 genes that only codify proteins. These genes are responsible for producing more than 100,000 proteins unique to the human body and they are all built using only 20 amino acids.

The twenty amino acids that make up our proteins are:

alanine
arginine
asparagine
aspartic acid
cistein
glutamic acid
glutamine
glycine
histidine
isoleucine
leucine
lysine
methionine

phenylalanine
proline
serine
treonine
tryptophan
tyrosine
valine

How do we obtain the amino acids to make our proteins? Well, the ones we cannot synthesize ourselves, we obtain by consuming other proteins. We disassemble them into amino acids and then arrange them in the sequence that we need. And since all proteins (plant or animal) use the same twenty, it doesn't matter where we get those building blocks, be it from a bean sprout or the rib of a cow. I have repeated this in many different ways to leave you without a doubt that the original protein is made by plants, from where all other proteins are derived.

All animal proteins are very similar. We have similar organs, such as skin, heart, eyes, ovaries, and digestive enzymes. This similarity makes it possible to accept organs from a pig, for example. It is how it was possible to use porcine insulin or thyroid hormone from pigs before we could synthesize these in the laboratory. It is also what allows us to use equine estrogen. However as you will see, the estrogens produced by plants (phytoestrogens) do not share a biological function comparable to the animal estrogen.

SCIENTIFIC CURIOSITY

What happens to the dietary animal proteins before they are completely broken down into the amino acid building blocks, when the portion still retains characteristics of the original protein? Do they interact with our human organism? This is exactly what T. Colin Campbell, PhD and his colleagues investigated.

Working for Cornell University, these investigators observed that casein (an animal protein) was capable of stimulating the multiplication of cancerous cells. While protein of vegetable origin (soy or wheat) did not have the same effect. Campbell stopped blindly promoting animal protein and dedicated himself to the investigation of this surprising finding. Curiosity is essential for science to advance.

THE PROTEIN CULT

Nonetheless, they encountered a problem. Already by mid-twentieth century there existed a protein cult that had become a cornerstone of the powerful U.S economy. Merely suggesting that animal protein ought to be reevaluated was an intolerable defiance. It unleashed a storm akin to the one during the era of Galileo in the 1600s—sentenced to lifetime house arrest for revealing that it is the Earth that circles the Sun. I will not recreate the saga to which these investigators were subjected, but it was brutal and persistent. This shameful chapter in modern science is detailed in *The China Study*, 2nd edition 2016 and in the books *Whole* and *The Future of Nutrition*, all by T. Colin Campbell, PhD.

The researchers of *The China Study* might not have had much money or influence, but they had plenty of integrity. They did not keep their findings secret and did not select which to publish or when. They shared their data with the world. What motivates someone to go against the accepted ways and risk it all to follow a smoking trail?

T. Colin Campbell, PhD, example of integrity and author of *The China Study*; he coined the phrase "Whole Food Plant-Based" to describe the pattern of food based on whole plants.

The China Study, whose principal author is Dr. Campbell, was instrumental in letting us know that cancer and other chronic non communicable conditions, such as heart disease and diabetes, cluster around privileged people where there is an abundance of meat, chicken, milk, eggs, cheese, creams, and confections. This discovery shook the investigators so deeply that they published the raw data (before interpreting it) so that every scientist in the world could have access to it and could continue investigating in search of the truth.

Nonetheless, in spite of learning about these discoveries, the medical class in the U.S. preferred to ignore them. Instead of including this information in the curriculum of medical schools and educating the public to help them avoid disease, they deemed it more attractive to manage the symptoms of diseases already declared. This model could be practiced from the comfort of an office with access simply to a prescription pad and a supportive hospital, without having to get involved with the community.

The plan of waiting for symptoms to then manage them with drugs has the components of the motivational triad popularized by Doug Lisle, PhD: it requires little energy, it is very convenient, and also very lucrative. There is also a revolving door between private enterprise, the government, and academia which permits well prepared scientists to work for the best bidder on biased research. Some of the actors that joined this movement are the pharmaceutical industry, the supplement market, medical insurance companies, and food industries, entities that are beholden to investors and stockholders.

If we add the messages that appear in medical journals, radio, television, internet, and social media, all of which repeat that diseases are inevitable and that we should all have access to drugs in order to live when in fact most of us could expect to live not needing them, we begin to realize that our health has been for sale.

The T. Colin Campbell Center for Nutrition Studies and the Physicians Committee for Responsible Medicine are two not-for-profit organizations that conduct and publish clinical studies without attachment to financial interests. Find such organizations in your area and support them.

The false concept that protein is something exclusive of animals has taken root in many people's minds with the help of propagandist messages and willing salespeople. Say you go out to eat and place an order of rice (3 g of protein) and beans (7 g) with a quinoa salad (6 g), half cup of pasta (3 g), steamed broccoli (5 g) and 3 ounces of mushrooms (3 g), and for dessert chocolate tofu (7 g) and pistachios (4 g), all of which add up to 37 g of protein; the waiter will still ask "would you like to add protein to your order?"

My colleague Fabiola tells a story of four girlfriends who went to a restaurant when they were internal medicine residents. The waiter greeted them asking, coming by yourselves? Meaning, don't you come accompanied by a man protein?

THE PROTEIC DICTIONARY

Each one of our cells has in its nucleus a map with instructions on how to produce 100,000 proteins. We achieve this impressive number of proteins using only twenty amino acids.

In spite of having identified over 500 amino acids in nature, around four thousand million years ago all living beings got established with just 20!

The twenty amino acids in the genome of most living beings are alanine, arginine, asparagine, aspartate, cysteine, fenilalanine, glycine, glutamate, glutamine, histidine, isoleucine, leucine, lysine, methionine, proline, serine, tyrosine, treonine, tryptophan, and valine.

With the same twenty amino acids that make plant proteins, we make our human proteins. We simply change sequence, order, length, and add various other molecules to the chain. This is how the bodies of all animals grow. Are you amazed?

No matter how many proteins or amino acids we consume, we only make the amount we need. Proteins are used, reused, or discarded. It should be clear that we do not store proteins. If you compare someone's urinary excretion of nitrogen (N) with their dietary consumption of protein, you can determine whether the person is consuming excess protein, because they will have a positive nitrogen balance (they will be excreting nitrogen). However, dietary proteins come accompanied by fats, and those we can store in unlimited quantities.

Digested proteins convert the amine group into ammonia and then into urea which is excreted in the urine; the remaining carbon skeleton (C, H, O) is used to:

synthesize other proteins
obtain energy (ATP), or
synthesize glucose (gluconeogenesis) during a prolonged fast.

THE WORD DICTIONARY

The letters of the English language are: a, b, c, d, e, f, g, h, i, j, k, l, m, n, o, p, q, r, s, t, u, v, w, x, y, z.

With the 26 letters in our alphabet, we can spell over 600,000 words in English and another 600,000 in Spanish. We use some letters more frequently than others, double the l or sometimes pair the ch and so on.

If you know the word game Scrabble®, it does not have as many "z" tiles as it has "a" tiles because one letter is not used as often as the other.

The board game Scrabble® illustrates that with 26 letters we can spell 600,000 words.

The board game Scrabble® in English comes with one hundred tiles (two are blank and ninety-eight are printed). Among them there are: nine A; twelve E; nine I; six N; eight O; six R; four S; and only J; one K; and one Q. This distribution of the 26 letters of the alphabet allows us to construct over 600,000 words in our language. The only limiting factor during the game is the number of tiles in play, the size of the board, and the empty places where we can add more words.

The 12 notes of an octave allow us to compose infinite melodies.

The piano keyboard features eighty-eight keys grouped in octaves: fifty-two white notes and thirty-six black half notes. With just these notes we can compose an infinite number of melodies.

Coming back to proteins, with the twenty amino acids mentioned we construct more than 100,000 different proteins. Just as in the word game, we don't need to have the same amount of each amino acid to construct the proteins that we need.

Proteins are super important, without a doubt, but we can produce all our animal protein by consuming enough plants. Animal protein is not superior to plant protein any more than English would be considered superior to Spanish.

The human body requires fuel for its construction and maintenance. The digestible carbohydrates in the diet turn to sugar; this is the fuel the body uses to generate between 40% and 60% of the energy we use to maintain temperature and function to capacity. The little protein and fat you need daily is used to complete the rest of the energy required for your activities, to synthesize humors, repair worn tissues and cells, or to replace those you leave behind in bathtub, bedsheets, urine, and stools. The propaganda about protein has you convinced that you ought to consume much more than you need. Don't you feel scammed?

Humor me as I tell you this story. Suppose metal, plastic, rubber, and glass are the proteins found in your car, which once in your possession it is already built. You add gasoline or petrol to the tank, because it is what it needs to run. You very well know that unless it is to replace some broken part, you don't need to add more metal, plastic, rubber, or glass.

LECTINS

Just like there are fanatics of the animal protein, there are enemies of the lectins. Let's see. Lectins are toxic proteins that plants produce to protect themselves from plagues. If you consume large quantities of them, they will also harm you. We find these proteins mainly in beans, lentils, peas, soy, peanuts, and grains such as wheat. They resist or survive stomach acid and we don't easily digest them. The symptoms when lectins are ingested raw are nausea, vomit, diarrhea, stomach upset, bloating, and gas. We can avoid these symptoms by cooking the products that have lectins or eating smaller quantities.

Some supposed experts are circulating disinformation about lectins with the purpose of "protecting the public from sure harm." Really, protect us? These folks take advantage of our blind tendency to follow social media and are building their brand spreading partial information about these proteins while harming the reputation of beans.

We should investigate what we hear about lectins (or any food) to verify its truthfulness and utility. Consider this: lectins are water soluble and heat labile. This means that the problem is solved by soaking beans and discarding the liquid or by boiling them for ten minutes. Coincidentally, these are the methods we have used for millennia to cook beans, right?

Furthermore, in the moderate quantities that we consume beans (one to two cups a day) and as part of a varied diet, lectins even show beneficial antioxidant properties. By choosing to eliminate beans from the diet you would deprive yourself of food that is associated with less cardiovascular disease, less type 2 diabetes, better gastrointestinal health, and normal weight.

Besides, the lectins in potatoes and tomatoes are not poisonous and the levels of these hemagglutinins in quinoa are very low. Canned beans already cooked are safe to eat straight from the can. Beware that if your slow cooker does not boil, lectins in the beans might remain intact.

Nutrition is way too important for you to eat or not based on something you heard, without investigating its truthfulness.

References:
Lectin Activity in Commonly Consumed Plant-Based Foods: Calling for Method Harmonization and Risk Assessment. *Foods* 2021 Nov; 10(11): 2796.

Toxicity of Kidney Beans (Phaseolus Vulgaris) With Particular Reference to Lectins. *Journal of Plant Foods* Volume 4, 1982

Polyphenol-Rich Dry Common Beans (Phaseolus vulgaris L.) and Their Health Benefits. *International Journal of Molecular Sciences* 2017 Nov; 18(11): 2331.

FATS

Each of our cells requires a certain amount and quality of fat. We call essential fats those we must acquire from the diet and non-essential fats the ones the body can make. Published studies confirm that by following a whole plant diet without added oils we obtain all the essential fatty acids to produce the rest. Once we accept these truths, we just need to concentrate on eating the bountiful variety that the earth offers us, until we are satisfied.

DHA

A fatty acid of great interest is docosahexaenoic acid (DHA). This polyunsaturated fatty acid from the omega3 series has multiple functions in the development of the brain and in cardiovascular health. We can synthesize it starting with alpha linoleic acid (ALA) from plants which is an essential omega3 fatty acid found in the walnut and the seeds of flax, chia, and hemp.

The scientific community has been keen in elucidating whether we can produce sufficient quantities of DHA, since people who do not consume prefabricated DHA often have blood levels that are considered low. Here is the situation, laboratory tests only detect what is in the blood (not in tissues) and DHA is used in tissues. It appears that:

> The brain has mechanisms to conserve DHA. For example, if we are deficient in polyunsaturated omega-3 (ALA, DHA, EPA), we increase the biochemical re-exchange of arachidonic acid (ARA) and reduce the re-exchange of DHA.

Although the synthesis of DHA from ALA is considered inefficient because the process has a <1% conversion, it is estimated that the brain needs between 2.4 mg and 3.8 mg of DHA per day, and it is observed that even that low conversion rate can supply the brain with enough DHA.

The comparable prevalence of neurological diseases in vegans, vegetarians, and omnivores suggest that alterations in the metabolism of DHA don't necessarily manifest neurologically.

> Reference:
> Is docosahexaenoic acid synthesis from α-linolenic acid sufficient to supply the adult brain? *Progress in Lipid Research*, Vol 59, July 2015, pp 54-66

Special circumstances:

> Babies fed on formula without prefabricated DHA have concentrations of DHA in brain tissue that are lower than those of breast-feeding babies. This indicates that in special circumstances, such as during early growth and development, we require more DHA and that mother nature (breast milk) provides.

VITAMIN B_{12}

The bacteria that synthesize vitamin B_{12} live in the colon (large bowel), while the vitamin's absorption happens in the ilium (small intestine) which is proximal or higher. In other words, we cannot absorb the B_{12} vitamin that we make in our own intestines. When humans used to eat produce directly from the ground, these came contaminated with vitamin B_{12} made by bacteria in the ground or poop made in the intestines of other animals. Today, bacteria make the vitamin in a laboratory.

Stomach acid liberates the vitamin B_{12} from the food so it can be absorbed. The proton pump inhibitors used by 20% of the population for acid reflux symptoms, reduce the stomach acid and create conditions that indirectly cause deficiency of this vitamin even, although less, when used as supplement.

> Reference:
> Proton Pump Inhibitor and Histamine 2 Receptor Antagonist Use and Vitamin B_{12} Deficiency. *Journal of the American Medical Association* 2013;310(22):2435-2442.

In 1964, Dorothy Crowfoot Hodgkin was awarded the Nobel Prize in Chemistry for describing the structure of the vitamin B_{12}. Its industrial production was achieved in 1972 using a selection of microorganisms that ferment sugars and other organic products. Before these discoveries, chronic B_{12} deficiency, also known as pernicious anemia, was lethal.

This vitamin interacts intimately with the microbiota. It is essential for the development and maturation of blood cells, and for our neurological function. The commercially available dose of cyanocobalamin of between 500 mcg and 1,000 mcg (micrograms) is many times the amount we need, which is about 2.5 mcg a day, so we absorb what we need and supposedly discard the rest. The safe superior limit of the vitamin is an area of early research.

We store many fat-soluble substances, such as hormones, toxins, and vitamins and readily eliminate those that are miscible in water. However, in spite of this vitamin being water soluble and not fat soluble, we manage to store it in the liver and other organs like the kidneys. We can evaluate our stores of B_{12} by measuring the levels of methylmalonic acid (MMA) in blood.

> Reference:
> Vitamin B-12 and the Gastrointestinal Microbiome: A Systematic Review. *Advances in Nutrition* Volume 13, Issue 2, March 2022, Pages 530-558

CALCIUM

Calcium is an element, so it is neither created nor destroyed. We have recycled it for millions of years. We obtain this mineral by eating plants. During the first 20 years of human growth, we deposit around 1,400 grams (3 pounds) of calcium in our bones. If we were buried directly into the ground, we would return that calcium to the soil, like all other animals who do not insist on impermeable coffins.

Every plant or mushroom that grows in natural soil absorbs calcium through roots and mycorrhiza to incorporate it in their stems, leaves, seeds, and fruits. Plants and mushrooms are our most reliable source of calcium.

COW'S MILK

The milk industry keeps millions of cows crowded packed in enormous buildings without natural light. Artificially inseminates them in violent filthy cycles to suckle a commerce that is never weaned. The cow's own babies never receive milk; white, pasteurized, and refrigerated, it is sent to market for your calves at home.

Every female of the mammal class produces milk for her young until they can feed themselves. Cow's milk provides the best nutrition for a calf born of sixty pounds and that by year's end will be almost nine hundred. Cow's milk is not for colts, whales, rats, or humans. The milk of a species is for the young of that species. No animal grows at the same rhythm as another.

Perhaps you have asked yourself why the government's nutritional guide *MyPlate.gov* indicates that cow's milk is essential for humans and recommends that we drink three glasses daily? This "recommendation" survives every revision of the guide thanks to commercial pressure from the dairy industry and the submissive acceptance by the United States Department of Agriculture.

> Reference:
> USDA-loses-a-battle-in-war-on-diet-guides. *Washington Post* archives 2000/10/04

The main excuse used for why they recommend milk consumption is calcium, without ever showing the evidence of how a cow is able to produce calcium rich milk when all she eats are plants. It is never explained why after weaning, no animal requires milk to grow healthy. It is also never mentioned that the body stops producing lactase (the enzyme that digests lactose) once we begin eating the solid food of our species. Sharing this information would be very inconvenient for an industry that depends on disinformation.

The Childhood Nutrition Program demands that schools serve cow's milk at each and every one of their breakfast, lunch, and dinner. Schools obey without saying a word, otherwise they lose government subsidies. This federal law allows the industry to discharge the excess cow's milk it produces, when the logical solution would be to produce less—which would in turn have them lose the subsidies it receives from the government, ha! It is a chain of special interests created for the benefit of a few at the expense of the many who have power, but do not exercise it. Where do you aim your protests?

The same law that expects schools to serve milk establishes that if a mother decides to continue breastfeeding after the first year, she is allowed to use her breastmilk as a substitute for the milk of cows! Please, read this sentence again.

Regarding any recommendation, it does not matter whether they come from a government agency or your doctor, you should always ask for evidence and expose who might be benefitting. If we truly know what the body needs, we cease complaining about the price of milk and stop buying it. We don't need it. If you have studied and know that the calcium your bones require comes from plants, you confidently refuse the tainted free milk and request more beans. If you understand that there is great power in being able to choose natural food low in saturated fat, food that brings nutrition in a matrix of fiber, vitamins, antioxidants, phytochemicals, and minerals, you are free to create health one truth at a time.

The United States Department of Agriculture operates in conflict with itself. On one side, they hold the public's interest, on the other the interests of their investors. What does it do? Instead of protecting the public, it backs the industry. Learn the truth and disseminate it.

> Reference:
> *Food Buying Guide for Child Nutrition Programs.*

The *2020 Nutrition Guide for Americans* proclaims that cow's milk is an indispensable food and recommends that we consume three portions daily. For the indoctrinated public, milk is synonymous with calcium and the dairy industry takes advantage of this myth to push the recommendation. It is sold in liquid form or yogurt, fermented, and butter, in baby formula, as ice cream, and in its most irresistible configuration: cheese. It is sold with fat and lactose removed, with added sugars and chocolate, altered with many other flavors, and supplemented with vitamin D, all to make it more attractive, so you present no objection. Never in the recommendations is the truth mentioned—that milk is not necessary for any animal beyond its breastfeeding period (especially the milk from another species) or its association with disease. No mention of the fact that is has antibiotics, contaminants, and saturated fat.

For a newborn, the benefits conferred by his mother's milk are unquestionable. After all, milk is designed by nature, the perfect food to give her young the best possible start in life. Among its essential nutrients are carbohydrates, protein, fat, calcium, potassium, and growth hormone. Nonetheless, the health risks associated with milk when consumed past this initial stage in our lives are considerable. It is deficient in nutrients that are only found in plants and is rich in growth hormone that instigates the multiplication of suspicious cells. The propaganda emphasizes supposed benefits of consuming milk, but according to the study LACE (which followed 1,900 subjects for eleven months), even a modest consumption of milk is associated with significant elevated mortality from breast cancer, which is kind of obvious, since those cells are in active growth.

Reference:
Life After Cancer Epidemiology (LACE) Study: A cohort of early-stage breast cancer survivors (United States). *Cancer Causes & Control* July 2005

Another study followed 53,000 North American women initially without cancer over almost eight years and found an increased risk of breast cancer of as much as 80% in those who consumed a moderate amount of milk. The positive qualities that milk might have do not cancel its negative properties.

References:
Dairy, soy and risk of breast cancer: Those confounded milks. *International Journal of Epidemiology* 2020 Gary E. Fraser, MBChB, PhD

Kristi Funk https://www.facebook.com/reel/1005661387825610

We also should not ignore that utilizing animals (which we have no right) desensitizes us to the inherent violence of an industry that brings sentient beings to life only to exploit during their short lives to then murder them. Consuming their products makes us participants in their bloody machinery. The person who decides to be vegan and eats exclusively plants joins the cycle of life and will even help ameliorate some climate changes.

Reference:
Eating Our Way to Extinction. Gerard Bishop

MONEY LEAVES A TRAIL

The tactics used by the dairy industry are the same used (and still used now) by the tobacco industry in order to make us believe that their product is innocuous. They discredit impartial scientific information, flood the environment with their alternate reality, and keep the purse in the hand of very few. With our votes, we elect a government to watch over public interests, but legislators need to be watched lest they side with industry. In a democracy, we have privileges and we also have obligations.

Reference:
Inventing Conflicts of Interest: A History of Tobacco Industry Tactics. *American Journal of Public Health* 2012 Jan;102(1):63–71

Investigation of the toxic products that result from burning the tobacco leaf have revealed substances that cause genetic disruption and cancerous mutations. It is indisputable that cigarettes are associated with cancer. Yet, we continue to battle with an indecent industry that suppressed evidence for decades and protects its economic interests knowing the harm caused by their product.

It took 7,000 studies and untold deaths before the medical community arrived as a consensus regarding the connection between smoking and lung cancer. Today we have a heap of studies with similar evidence about animal products and their effects on our health. There have to be people willing to risk their investments in service of humanity. Let us not forget that industries are made of individuals. All that is needed is for some to raise their voices, for others to feel shame, for many to feel courageous.

In addition, if we viewed our eating conventions through an integral or wholistic lens instead of a reductionistic one, we would see how eating animals contaminates water, soil, and air. It depletes planetary resources that are not renewable during our lifetime, such as fresh water deposited in underground aquifers drop by drop through millennia; petroleum, the fossilized remains of our ancestors; and the forests, those ancient lungs that purify our air.

What are we waiting for? The change we want begins like the roots of a sapling, before the first leaves erupt, underground.

ANIMAL MEAT

Billions of animals are made annually in the United States so people can chew their meat. They are not brought to this world to live healthy lives, but to be corralled, penned en masse, medicated and artificially inseminated. They live confined. Brought up in filth. Kept until they reach a size when they can be transported to a slaughterhouse to be killed, bled, skinned, and quartered. The sole purpose of their lives is to be turned into money.

They are finally brought to market packaged in plastic, without a trace of who they were, so you don't recognize them. This dissociation enables you to love dogs, cats, and parakeets while you go on eating pigs, lambs, and chicks. My niece, whose heart is filled with love for abandoned animals, one day told me "Oh auntie, don't ask me to think about such things."

Maintaining this separation takes energy, but you persevere. You know that animal meat is not an ingredient, that it is the body of a sentient being, but you dissociate this fact to enjoy the next bite. The industry dedicated to killing animals is laborious and replete with physical and psychological risks for the worker. It is expensive to keep eyes and heart shut.

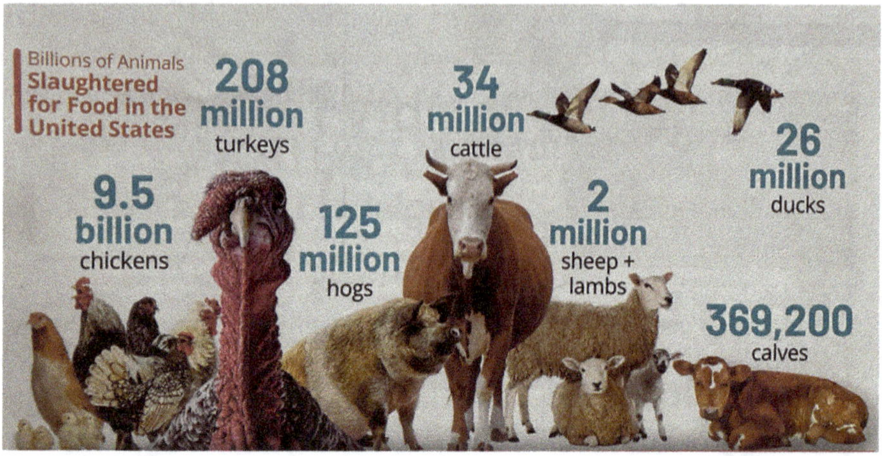

Every year, the United States kills billions of animals to consume their meat: more than 9 billion chickens, 208 million turkeys, 125 million pigs, 34 million cows, 2 million lambs and goats, 26 million ducks and 369,200 calves.

VEGANISM

Veganism is not a diet, it is a way of life. The vegan will not exploit other sentient beings for any purpose. Therefore, as much as possible, will avoid using animals for food, clothing, entertainment, in adornment, zoos, investigation, or education. Having said that, the idea is not to create an elite of "I am more moral than thou." The vegan lifestyle opens the heart, does not shut it. We walk imperfectly with dignity. We lend each other a hand. So what if it takes us years? Compassion can be contagious. All for one, and one for all.

The definition of veganism according to the Vegan Society remains loyal to the vision of the founding members, just as we work toward a world where humans do not exploit other animals. We are more committed than ever to promote vegan lifestyles for the benefit of all animals, people, and the environment.

Reference:
The Vegan Society

It used to be considered normal to torture and sacrifice animals in order to test for the safety of cleaning or beauty products, to teach anatomy in schools, or to train medical and rescue personnel, it was the norm. The attitude that informs such behavior is under revision. Today, we have many materials, products, and tools which serve those purposes. We have electronic mannequins and standardized technologies being used in schools and training. Speaking in strictly practical terms, we don't need to think of using animals for anything.

In April 2025, the commissioner of the Food and Drug Administration (FDA) announced the gradual elimination of animals in the production of monoclonal antibodies with the intention of extending this to all drugs tested on animals. The Physicians Committee for Responsible Medicine is committed to phasing out the use of animals in every area to the extent possible. I know of no similar organization.

EGGS

The egg industry follows a similar trajectory to the meat and dairy. For this industry, the bodies of designated "laying" hens have no other purpose but to lay eggs. What is important is that they arrive clean and in boxes without a trace of blood or suffering, so your heart won't know the truth. The funny thing is that, although unconscious, you do perceive it. We might ignore the details, but we are not innocent.

Defenders of disinformation: a renowned physician says "eat more eggs, because they are powerful antioxidants and contribute to the retina health." Eat more eggs because they have carotenoids! You did not use to eat eggs, but since they are constantly making the news on the internet, you now consider adding them to the menu. Yes, eggs have carotenoids. They come from the hen's plant diet. This information is so reductionistic that it invalidates itself. Eggs have carotenoids, but lack carbohydrates, fiber, phyto chemicals, and vitamin C. Plus they are chuck-full of cholesterol. These orange pigments called carotenoids come better packed in green, red, yellow, and orange plants that we should be consuming regularly in quantity and variety. Which way do you lean?

Eggs in context: Contrast the following investigation which considers eggs in the context of the American diet, where overnutrition, overweight, and obesity reign. The study concludes that the increase in cholesterol attributed to the eggs could adversely affect cardiovascular disease prevention and contribute to premature death. Remember that heart attacks are the number one cause of death. Reductionism is myopic.

Reference:
Associations of Dietary Cholesterol or Egg Consumption With Incident Cardiovascular Disease and Mortality. *Journal of the American Medical Association* 2019;321(11):1081-1095.

OIL

Let us now talk about oil, since after milk and protein it is the most discussed ingredient. Among its consumers, olive oil enjoys singular status. It has surnames such as "extra virgin," "virgin," "cold pressed," "pure," "light," and "olive pomace." In spite of leaving a layer of sticky grease on stove, pans, and walls, for many cooks oil is gold and they protect it with their lives. I dare bet that even if trying to cut back on fat, you consider olive oil an indispensable ingredient.

The Forks Over Knives documentary brought to light the significant studies of Caldwell Esselstyn on heart disease.

YOUR REPUTATION FOLLOWS YOU

The 2011 documentary Forks Over Knives made public the benefits of a diet without added oil for people with cardiovascular disease. However, today the organization subscribes to the use of olive oil as an optional ingredient. To accept this relaxed position in the use of the olive oil as "new information" that dethrones prior assertions about the effect of oil upon the endothelium and blood flow, is to omit its historic context.

It has been proven that in the short term, olive oil reduces arterial blood flow more than other oils, meaning that less oxygen and nutrients flow to tissues after the fatty meal. Not exactly what you associate with the royal Mediterranean diet, right? Studies continue to investigate if there is a long-term health reason to recommend the use of olive oil and which version of the oil takes the prize. To date we are still accumulating data.

Reference:
Effect of Extra Virgin Olive Oil on Anthropometric Indices, Inflammatory and Cardiometabolic Markers: A Systematic Review and Meta-Analysis of Randomized Clinical Trials. *The Journal of Nutrition* 2024 Jan;154(1):95-120

Every oil is extracted from a plant in a process that discards water, fiber, protein, carbohydrates, vitamins, and a host of other valuable phytochemicals. As such, the oil is no longer accompanied by the other musicians of the band. It comes alone, 100% fat. It might be able to realize some functions by itself, such as cross the cellular membrane and participate in the elaboration of some biochemical compound, but it comes isolated and in concentrations unfamiliar to the body. Meaning, without the buffering protection which the body expects of other ingredients in the whole plant.

Oil is the fraction of fat from a fruit (olive), grain (corn), nut (walnut), or seed (sesame, flax). The calorie density of oil is 4,000 calories per pound. Compare this with spinach that contains only 100 calories per pound.

To obtain a tablespoon of olive oil, one must squeeze between fifty and one hundred olives. Whether from the olive or any other fruit or seed, oil renders 4,000 calories per pound. Compare this with one pound of cooked lentils at only 1,300 calories and an enormous amount of nutrients. Remember this concept of calorie density for when you are stuck in overweight.

If you have been diagnosed with fatty liver, prediabetes or diabetes, you will be required to consume no more than thirty grams of fat a day to regain insulin sensitivity. This is exceedingly difficult if you insist or persist in the use of oil. One tablespoon of any oil, including olive oil, is fourteen grams of fat.

Every whole plant food comes with a natural fat component of pretty much what the body needs relative to the other macronutrients (carbohydrates and proteins). When you use bottled oil, you take the equation out of proportion. If you are intimidated by the prospect of cooking without added oil, it's easier than you think.

The Center for Nutrition Studies has published over seven hundred plant-based oil free recipes in English and in Spanish.

References:

Olive, soybean and palm oils intake have a similar acute detrimental effect over the endothelial function in healthy young subjects. *Nutrition, Metabolism, and Cardiovascular Diseases* 2007 Jan;17(1):50-7.

The postprandial effect of components of the mediterranean diet on endothelial function. *Journal of the American College of Cardiology*, Vol 36, Issue 5, 1 November 2000, 1455-1460

Role of free fatty acids in endothelial dysfunction. *Journal of Biomedical Science* volume 24, Article number: 50 (2017)

A BALANCED DIET

Following a balanced diet is something many profess and too many misinterpret. They think that to follow a balanced diet means to eat in moderation, by which they mean a little of everything. In truth, to follow a balanced diet means to consume a great variety of whole plants of many colors every day. More details are included in Part Six.

A balanced diet consists of an abundance of leafy greens, fruits, vegetables, grains, beans, mushrooms, nuts, and seeds. This diet satisfies our nutrition and is for everyone.

CALORIES

A calorie is not a nutrient, but a unit of energy. Foods store energy in what we call calories. A calorie is equivalent to the heat necessary to raise the temperature of a gram of water by one degree Celsius.

Just like the batteries, foods concentrate energy.

CONCENTRATED ENERGY

When we speak of calories in food, we refer to the energy that can be obtained or liberated through the combustion of its macronutrients. The three macronutrients in food and the calories per gram of each are as follows: proteins (4 cal), carbohydrates (4 cal), and fats (9 cal). In addition, the microbiota liberates 2 calories per gram of fiber and while alcohol is not considered a nutrient, we recover 7 calories per gram of digested alcohol.

It is not necessary to be obsessed with the amount of calories you consume each day, but it is indeed worthy to have an idea of the energy stored in foods, as a matter of general knowledge or food literacy. It is of value to know how the calories in leafy greens, vegetables, fruits, grains, beans, mushrooms, and nuts compare with the calories in red meat, chicken, milk, eggs, cheese, oil, pastry, fritters, and other fast food.

Calorie density (calories per weight):

> one pound of leafy greens, such as spinach or kale=100 calories
> one pound of olive oil=4,000 calories
> one pound of fruits=250 calories
> one pound of cheese=1,800 calories

I invite you to create a table like this with some of your favorite foods. Take the time to do it. This task will improve your knowledge of the foods you consume. Find this basic information easily on the internet or in an encyclopedia such as *Wellness Foods A To Z*, published by the University of California Berkeley in 2002, available second hand. Although some information on recommendations is antiquated and plain wrong, the nutrition labels are standard scientific values and have not changed.

For a diet to be sustainable, it needs to provide enough energy. If you only eat spinach, you would have to ingest 20 pounds (9 kg) to arrive at the average 2,000 calories. If you eat one pound of nuts, you will reach 2,000 calories, however you will be deprived of an enormous diversity of nutrients. Once you assimilate these concepts, eating a balanced diet will not require stopping to weigh or measure your foods.

THE HORSE

I will speak of the horse, so we can better understand ourselves. The horse is such a strong animal that we classify big machines in terms of horsepower. This animal consumes all its calories in plant form. The mare gives birth to a colt of 100 pounds (45 kg) and rears him on her milk which she makes eating only plants. (One pound equals 0.45 kg.)

The calories that a horse needs:

A horse of 1,000 pounds (453 kg) must consume 16,000 calories per day simply to maintain his body.

Of this total, about 2,268 calories should come in the form of protein.

The hay (cut grasses dried for fodder) that he eats provides 220 calories per pound. Therefore, the horse must chew and swallow 22 pounds of hay to obtain the proportion of nutrients he needs.

Of the 16,000 total calories, 12,000 should be carbohydrates which the horse will convert to glucose and burn to generate his energy.

As you well imagine, hay carries protein, carbohydrates, fat, micronutrients, fiber, vitamins, phytochemicals, antioxidants, and water. All these nutrients are necessary for the functioning of the horse's body.

THE HUMAN BODY

The calories that a human body needs:

In comparison, a sedentary human of 150 pounds (68 kg) requires about 2,000 calories.

The calorie density of our food is varied: 116 calories of protein per pound of cooked beans, 80 calories per pound of cooked grains, 60 calories in a pound of vegetables, 12 calories per pound of fruits, 60 calories in a pound of mushrooms and 360 calories per pound of nuts and seeds.

Of the 2,000 calories per day, we need 1,040 to 1,600 in the form of carbohydrates, since just like the horse, this is the nutrient that provides us with glucose for energy. Therefore, with two to three pounds of food we can satisfy our protein requirements and the proportion of nutrients that we need.

Do you feel better informed?

CALORIES COURTESY OF THE FOOD INDUSTRY

The food industry launches 4,000 calories per person per day. This is twice the amount that an adult needs. In commercial terms, this merchandise must be sold in order to have net earnings. They lure us with offers of two for one, larger portions, make their products sweeter or saltier, add addictive ingredients. They ultraprocess their product to win your palate over and outperform the competition which is engaged in the same game. In our economic system, industries do not adjust by producing less, they instead fight to sell their excess to consumers who keep on buying (or not). In which camp are you?

WHOLE AND PROCESSED FOOD

A "whole" food is intact, just as it was harvested out of the ground. Even when you chop, freeze, dehydrate, grind, mash, blend, boil, heat or condiment a whole food, it is still a whole food. The unhulled ground sesame seed that we call tahini; the whole wheat grain or whole chickpea that we grind and turn into a flour; the seed or peanut that we grind to a buttery paste, and the ground date which we turn into a sweetening paste are all versions of foods still "whole."

Conversely, a "processed" food represents a fraction of the total, a portion that was taken out of the whole. White rice is processed brown rice. It is just the endosperm of the grain, since in the process the germ and the bran were both discarded. The same way, to make corn oil, the fat is extracted out of the whole fiber, water, protein, carbohydrates, vitamins, and other phytochemicals. Table sugar is the saccharose (glucose and fructose) of the cane or the beet—a processed or refined carbohydrate with absolutely nothing that identifies its original source.

As a curious and pertinent note, the tools used to squeeze oil from grains and seeds were invented around the year 3000BC; the sugar refinery plants appeared by the year 400; and the machinery to extract sugar from corn were invented in the twentieth century.

To extract components or fractions out of grains, fruits, or seeds requires a process that is outside our reach at home. If you want to obtain one tablespoon of oil, you must press between fifty and one hundred olives then discard the fiber, protein, carbohydrates, and other nutrients of the fruit. We wouldn't dream of filling oil bottles at that pace in our kitchens.

Similarly, consider the industrial process of extracting sugar from beets or making sugar out of the starch of maize.

The purpose of presenting this argument is not to convince anyone to not use the products of modernity, but to highlight that we have replaced the original sources of whole nutrition, whole plants with which our digestive and immune systems evolved, and threw ourself head first to use sugars and oils for everything and in everything.

We could more or less begin this process of separating some foods in our own kitchen. For example, mashed fruits or ground nuts and seeds tend to render a pulp with a liquid. Since the water content of fruits is so high (95%) and also the oily content of nuts and seeds (80%), as the fiber breaks, the components separate almost spontaneously. Mix them together before using them and you still have a whole food. You may have noticed when you open some commercial peanut butter jars, that much of its oil floats above the paste.

I invite you to grind half a cup of nuts until you no longer see intact pieces and let it rest for a few hours. You will see a cloudy oil layer forming over the paste. Likewise, cut a pineapple in small chunks and you will soon find it in a pool of sweet juice. If you only use the oily portion or the juice portion, you are using a processed part of the whole food. With these examples it should be clear what is a whole food and a processed food.

Lastly, the term "elaborated" refers to the action of combining various ingredients and spices to create a new dish. We can elaborate endless delicious dishes using whole plants or processed ingredients. The more whole ingredients we use, the healthier our elaborated dishes will be.

Food factories elaborate flavor bombs mostly with processed ingredients.

FACTORY FOOD

Most factory food is elaborated with processed ingredients extracted from the whole (flours, oils, sugars, and salt) to which other ingredients are added: be they ultra processed with superpalatable addictive properties or of non-plant origin (such as eggs and dairy). Their attractive packages take up the largest space in any supermarket and they last for months on the shelves—since they lack nutrients to attract even molds, bacteria, or insects. However, we dear humans seem to fall for the artificial colors, smells, and tastes.

Industries shamelessly seek the youngest palate to hook them early. After a child has tried this manipulation masqueraded as food and likes it, the company can rest on its laurels. Now it is you who must deal with the tantrums of the young addict, who does not want to try the beans you put effort to make for dinner.

Food factories are dedicated to elaborate infinite irresistible combinations never seen in nature. Do you still wonder why there are so many people with obesity or why organizations like Overeaters Anonymous have so many members?

OUR BIOLOGICAL PATTERN, SCHEME, OR WIRING

Every cell of the body, which in an adult is around 37 trillion, needs nutrients and oxygen to function.

Our health depends on thousands of simultaneous exquisitely orchestrated functions; none can be completed independent of others or without affecting the others. An organism who has at its disposal a variety and an abundance of nutrients, will use them expertly without tripping. For optimal functioning, we need that almost 80% of the calories be available in the form of carbohydrates. These will be turned into glucose which we use for energy to the tune of four calories per gram.

We complete the 100% with 10% in the form of proteins and another 10% in the form of fats. These nutrients are used to help us grow, repair cells, produce hormones and enzymes, replace spent substances, fabricate muscle, and substitute the cells we leave behind in our environment. With these nutrients we can also generate energy (four and nine calories per gram of protein and fat respectively).

Proteins are not stored, but fat is, as you can observe all around. The adipocytes are lifesaving cells that store fat so we can survive the times when fruits, vegetables, mushrooms, grains, seeds, and roots are scarce. We evolved capable of storing fat to survive. The other cells can also store fat, but never had to deal with an excess. We never had to put the brakes on storing fat, until now.

Obesity, inflammation, atherosclerosis, and insulin resistance, among others, are consequences of a limitless storing of fat everywhere in the body. What used to be a lifesaving floating device is now a block of cement.

We have changed our environment much faster than our capacity to deal with it. We walk the supermarket picking up food for the refrigerator and pantry as if we were roaming the prairie. With laser eyes we search for what can be turned into sugar for quick energy and fat rich food that we can store for later. We have been practicing this survival routine for millennia. The problem is that we no longer live in the Stone Age, not even in the Twentieth Century. We live surrounded by products that have been elaborated with superpalatable combinations of sugar and fat that do not exist in nature with no protection, because genetic evolution will take too long to provide us with any.

The adage goes forewarned is forearmed. However, in this case to have knowledge is not enough, since we must impose limits on our behavior and those limits go against our biological wiring. To force ourselves to consume less sugar and fat, when they are within reach, is very difficult because our instincts operate unconsciously as reflexes. To refuse the fatal but attractive combination of sugar and fat in a food makes some people tremble. I insist that we are capable, because we are not weaklings. Having said that, to succeed we need a serious plan and to follow it for long enough so that it becomes a new routine. This requires help, support, and community.

The most effective strategy I have seen people use consists of:

1. Eliminating from the surroundings all problematic food (home, work, routes).
2. Avoid exposing oneself to uncontrollable situations (insistent people and events)
3. Implementing the plan for success without exceptions.

My patients say "It will be difficult, but not impossible."

The intention of this project "We Create Health One Truth at a Time" is to equip you with truths so you can make good decisions. With a solid foundation, you can keep your eye on the goal and focus on the prize which is to feel well and be healthy.

THE PLEASURE TRAP BY D. LISLE AND A. GOLDHAMER

According to a theory supported by Doug Lisle, PhD, of True North, human beings look to satisfy their needs and desires with a minimum of effort or trouble. This is the same triad that channeled medicine to the treatment of symptoms, instead of adopting the difficult task of disease prevention. The medical-industrial complex knows very well that switching habits costs energy, time, and work, instead it offers us an easy facsimile of health in a pill. And we say, why not?

However, when we make such a pact of convenience, we welcome much more than a pill. We open the way to a plethora of drugs and their side effects, to obesity, hardening of the arteries, chronic inflammation, insulin resistance, cancers, and to auto destructive and neurodegenerative conditions, such as Alzheimer's for example.

Medicine paves this way in collaboration with the pharmaceutical and food industries. We are barely recognizing that 80% of our present-day ailments are conditions we could avoid and even revert. It is not too late to wake up to the reality and act in service of your health.

Once we get used to food that is unnaturally and excessively sweet, salty, or full of fat, it is hard to let go. Asking the palate a return to whole foods requires an effort akin to stopping a drug. See if you recognize this situation:

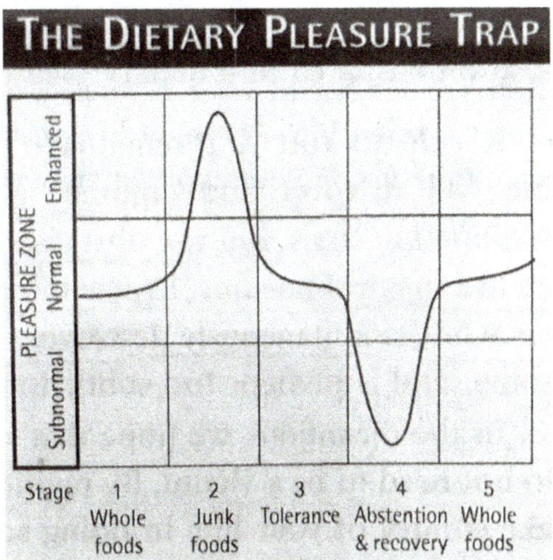

The Pleasure Trap: easy to fall into it, much harder to crawl out.

The Pleasure Trap Trajectory

You were innocent:

You liked apples and strawberries.
You enjoyed cracking, peeling, and eating raw walnuts and pistachios.
Baked potatoes filled with well-seasoned black beans were your weakness.

One good day:

You tasted a caramel coated apple and strawberries dipped in chocolate.
You received a bag of salted toasted nuts.
You put your hand into a bag of salty fried potato chips.

Will you become instantly addicted?

Of course not, but you felt the pleasant dopamine hit and if you kept these treats handy, they would become routine. Every time the pleasure is somewhat less intense, but enough to keep you coming back for more. You get used to it, even as the pleasure diminishes.

You fell into the trap:

You lost the initial exaggerated sensation of delight. But you haven't lost your head, you still remember that the natural flavor of your food was delicious and the sensation very enjoyable, plus it was less expensive. You decide to return to your apple, strawberry, walnut, baked potato stuffed with frijoles.

Abstinence syndrome:

There is one problem, you no longer have a taste for the food you used to like. Having adjusted to the artificial exaggerated combination of sweet/salt/fat, now your taste buds register negative pleasure with the unadulterated food. You try for a few days, but no way, so you throw in the towel. Oddly, in the pleasure trap one feels no pleasure. You need the charmed food, not to feel pleasure, but to quench a desire. This is known as the abstinence syndrome.

Determination:

This is where many people get stuck, like in quicksand. The more they try, the more they sink. The sensation of impotence and frustration is intolerable. Getting out of the trap will take earnest insistence, resolve, and help as one drags through anguish, malaise, and pain—this is no exaggeration.

It is not easy:

I have seen hospitalized persons pry themselves off the claws of heroin addiction. It is painful to witness the sweats, the trembling, the fevers, and even the convulsions of someone fighting to come afloat. It is not the same with addictive food, but the experience is intense and I don't want to minimize it.

The prize:

You will feel anxious thinking that the drug will be out of reach, that you won't be sinking your teeth into a piece of fried meat, that your pizza won't be covered in cheese, that your morning coffee will come without pastry. The change will be difficult, but the prize is life itself.

TRADITIONS AND HABITS

Why do we change a behavior or a habit? Two powerful reasons make us change behavior: one is when we see a better future than the one that awaits if we continue the current trajectory; and another when the current habits no longer serve us well. In other words, we change because we experience an internal shift, not because someone convinces us to change.

Humans have arrived at this evolutionary moment having consumed ingredients as they come out of the earth with the enzymes to digest them, Our bodies are intimately familiar with garlic, onion, beet, asparagus, cauliflower, lemons, pumpkin, squash, potato, corn, peanuts, dates, coconut, rice, and beans. Our intestines house microbes that feed on the fiber we cannot break ourselves. These unicellular organisms are permanent residents that contribute to our nutrition with their valuable discarded metabolic products or postbiotics (their "poop"). This elegant coevolution has taken one hundred thousand years. We have prospered as a species eating mostly delicious whole plants.

Our scientific curiosity has facilitated the fractioning of earth products to separate them into their components and render us flours, sugars, and oils. At face value, this is a valuable scientific achievement, but we began playing with these fractions and created edibles that have never existed in the natural world. Now what?

Take for example this recipe: 4 cups of flour, 1 cup of sugar, 1/2 cup of oil, milk from a cow, eggs, salt, spices, and pumpkin puree. We've combined some whole ingredients with cheap partial ingredients, we oil the corporate machine, and we hand the palate a super product unknown by our system. Out of this combination we obtain not a pumpkin pie, but a pie with a little pumpkin.

This type of food poses challenges to our system, since unlike the grain, vegetable, or seed, it does not arrive enveloped in much fiber, has none of the phytochemical components that usually work together, and lacks the proportional balance among nutrients. Absent all this and once metabolized, the body will store the fat, use the protein to repair tissues, and will push the little fiber or indigestible carbohydrate to the colon, where with some luck lives a population of hungry bacteria that can produce some postbiotic of value, and you survive.

This nourishment is of mediocre quality. In time, the scale tips toward excess triglycerides that little by little will fill adipose cells and then fill other cells in the body. This pattern easily puts us on the path to develop insulin resistance, prediabetes, high blood pressure, and elevated cholesterol, deteriorating our health which is so valuable.

The body metabolizes processed and ultra processed food much easier than food in its whole or integral state, we could say that it comes already chewed and digested (or decomposed). Unlike from a sweet potato, we absorb much more energy from a donut that has no protective fiber and gives us naked nutrients in the form of milk, eggs, excess flour, oil, and sugar. This way of receiving nutrients is not beneficial, we can see it reflected in the rates of obesity and disease all around.

In order to successfully plan a healthy and sustainable diet, we ought to take into account the caloric and nutritive value of food. Remember that the avalanche of food coming out to the market daily is 4,000 calories per person (who needs that much?): there is no better way to obtain nutrition than from whole plants.

FREE HUMAN BEINGS ARE UNPREDICTABLE

It is quite possible that you already have all the information you need to change your ways, but you lack a serious motivation. A person changes because they want to arrive at a different future, they want to be in shape to climb Machu Picchu, or perhaps because the old habits no longer serve them well, alcohol is ruining the relationship with a loved one, or they are out of breath playing golf. Some years ago, I went to the public pool twice a week for six months to develop the confidence to swim around an islet in Hawaii.

We must be equipped with truths, but the decision is not purely intellectual. There could be an intense desire to make a change and at the same time an external force creating an obstacle or resistance that slows down positive change. Something must touch our deep internal fibers. Change requires consciousness, earnestness, and perseverance. If you are trying to change someone else, redirect your energies. Remember that your most important project and with the only one you could be successful is you. Change does not happen by willing it on someone else.

Reference:
Postprandial energy expenditure in whole-food and processed-food meals: implications for daily energy expenditure. *Food and Nutrition Research* 2010; 54:5144

EYES OF AN EAGLE AND EYES OF A MOUSE

We can only appreciate the forest from afar, with the eyes of an eagle. From the ground, from the point of view of the mouse, we observe the forest one tree at a time.

THE PAPAYA

Say one day we come across a delicious green fruit, similar to a pear, but enormous with intensely orange pulp and black seeds like caviar. It grows hanging from a tropical tree that in only one year reached twenty feet (six meters) tall. We study it to get to know it deeply and find that one cup of the such papaya packs 55 calories, one gram of protein, 14g of carbohydrates, 0.2g of fat and provides 2.5g of fiber, 360g of potassium, 4mg of sodium, 87mg of vitamin C and 53mcg of folate. The fruit also has bioactive components (carpaines, benzyl isothiocyanate, latex, papaine, zeaxanthine, choline, etc.), precursors of vitamin A, as well as vitamins B, C, E, and K, pantothenic acid, lycopenes, lutein, magnesium, copper, and calcium. If that was not enough, it has plenty of fiber and antioxidants, which are associated with digestive benefits and neutraceutics. Impressive.

It seems like a good idea to separate the components in the lab and bottle each fraction in separate pills. They could last months or years, potentiate our antioxidant capacity and simultaneously turn into a lucrative business. The pharmaceutical industry already serves the symptomatic public, but who will serve the healthy neurotic public that has no symptoms? Enter stage: the supplements. What a great idea!

The economy: you can buy two papayas a month for $120 a year. Although buying papaya supplements would equally cost $120 a year, the supplement market could be more profitable, since it does not handle, transport, or dispatch a perishable fruit.

SUPPLEMENTS

Here is a problem with respect to supplements in our bodies. Investigations demonstrate that unaccompanied by the other nutrients in the whole fruit, these fractions rarely function the way we anticipate or behave the way we wish they did. Each fraction works within the web of the fruit as a collaborating part of the unit. Nonetheless, this fact has never been an obstacle for an industry that operates without regulation and that, disregarding this reality, will continue to line up their products on the shelves of food shops and drug stores.

Without the regulation within which the pharmaceutical drugs operate, the supplement industry can attribute any number of unsubstantiated benefits to its products. For example: that they reduce cholesterol in arteries, prevent arthritis, slow aging, fight cancer, protect against macular degeneration, reduce the risk of cardiovascular disease, alleviate stress, increase platelet count, control dengue fever, facilitate digestion, guarantee weight loss, and are backed by many professionals. Anything else? These unfounded promises have had success attracting a public that prefers to imitate health rather than work to create true health one bite at a time.

The papaya is a papaya only in its integral version. It stops being a papaya when we separate it into enzymes, proteins, carbohydrates, fats, vitamins, phytochemicals, antioxidants, fiber, and water. By itself, one component is incapable of completing biochemical reactions that require the interaction with other components. Nutrition results from the exquisite and elegant collaboration among the nutritive foods you give the body.

If we use the eyes of a mouse, we lose the "whole" perspective. When we isolate a component and elevate it above the whole, the investigation loses its applicability.

A papaya will never be represented by the carotenes in the fruit or by how much fiber or vitamin C it stores. It will neither ever be identified as an antioxidant molecule or a hydrogen donor. That fruit we call papaya is much more than simply the sum of its parts.

However, in spite of science showing that the bioactive components do not function in isolation, many people are convinced that it is more sophisticated to spend money on supplements rather than eat whole plants which are within the reach of every plebeian. Are you spending your paycheck on pills rather than food?

It is estimated that the per capita expenses in fruits and vegetables in the United States is between $200 and $500 per year. The amount spent in vitamins, shakes, and dietetic supplements is between $100 and $500. Wow! This industry has captured those interested in a facsimile of health, more than true health.

The work of dismantling myths regarding nutrition needs that each of us add our efforts. Do not ignore the references included here, every one of them supports the text and the importance of giving the body what it needs. Find the way to check them out.

"We are going to lose our democracy unless Americans, and particularly young people, take steps to educate themselves and combat the misinformation chaos created by the rise of the internet." —Sonia Sotomayor, Justice of the Supreme Court of the United States.

THE FACES OF REDUCTIONISM

With technological advances, it has been possible to isolate individual components of plants. For example, fiber became an economic entity when medicine began using it to treat constipation. Today it is sold separate from plants as if that was the best way to get it, often replacing the consumption of healthy whole plants. In fact, you well know that all plants make these resistant carbohydrates where they are accompanied by valuable components found only in the whole plant. So, I am sure that you get your fiber by eating oats, mango, pumpkin, and cauliflower and not in pills or powders, right? We use the generic term fiber for all resistant carbohydrates, but we should take advantage of the many varieties of fiber in fruits, grains, beans, mushrooms, and vegetables. A well-informed human being is worth so much more than a web of propaganda.

References:
Fruit, Vegetables, Dietary Fiber, and Risk of Colorectal Cancer. *Journal of the National Cancer Institute*, Volume 93, Issue 7, 4 April

2001, p 525–533
Dietary fiber intake and risk of colorectal cancer and incident and recurrent adenoma in the Prostate, Lung, Colorectal, and Ovarian Cancer Screening Trial. *American Journal of Clinical Nutrition* 2015 Oct; 102(4): 881–890.

Potassium is another example of misplaced attention on one ingredient, an essential nutrient for cellular function and which the public associates exclusively with bananas. This association is without merit, since potassium is abundant in many plants, such as beans, mushrooms, and potatoes. However, ask anyone to tell you where do we get potassium. The banana industry benefits from our misinformation, so it neither complains nor corrects us.

One more example which we already discussed is calcium—it passes from the soil to the plant then to the stomach of animals, who incorporate it into their bones, teeth, and tissues to eventually once dead liberate it again to the soil, from where it is absorbed by the plants, which animals then eat and so on, ad infinitus.

Many physicians ignorant in nutrition (but educated by the cattle industry) fail to recognize this biological concept. Instead of promoting the consumption of plants, they dare denigrate mothers who do not give cow's milk to their children. The system is not broken, the system is working very well for those it benefits. I emphasize that cow's milk (and that of every mammal) is rich in calcium, because the animal eats plants. In the case of carnivores, they must eat the bones of their herbivore prey who ate the calcium rich plants.

In the nutrition guide, the Department of Agriculture recommends drinking three glasses of milk a day. The medico-pharmaceutical complex elaborates commercial baby formulas using cow's milk. Women come out of their bone densitometry appointment string cheese in hand as a reminder of their dependable calcium source. (Only 6% of these tests are realized in men.) With this project, we are unmasking the system and protecting us from falling in its traps.

PART THREE

Diseases of the Day

Since 1950 there has been an explosion of obesity, diabetes, hypertension, atherosclerosis, Alzheimer's, cancers, and autoimmune diseases; all are non-communicable diseases, that is to say, they are not contagious or transmitted like influenza. They spread through dietary habits, they run through the kitchen.

The body mass index in adults (kg/m2) is commonly represented in a tri-colored graph, with height on the horizontal axis and weight on the vertical axis. The green band represents the "normal" weight index for each height; the pink represents "obese" weight, and the lower yellow band represents "under" weight. The middle yellow segment between the green and the pink represents "over" weight.

I use this graph to compare a person with herself at different times in her life. How much did you weigh at age 20 and how much do you weigh now? If when you stopped growing tall you carried a normal weight, you should stay on that band for the rest of your life. Many young people today, already overweight in elementary school, will have no normal weight with which to compare themselves come age 20.

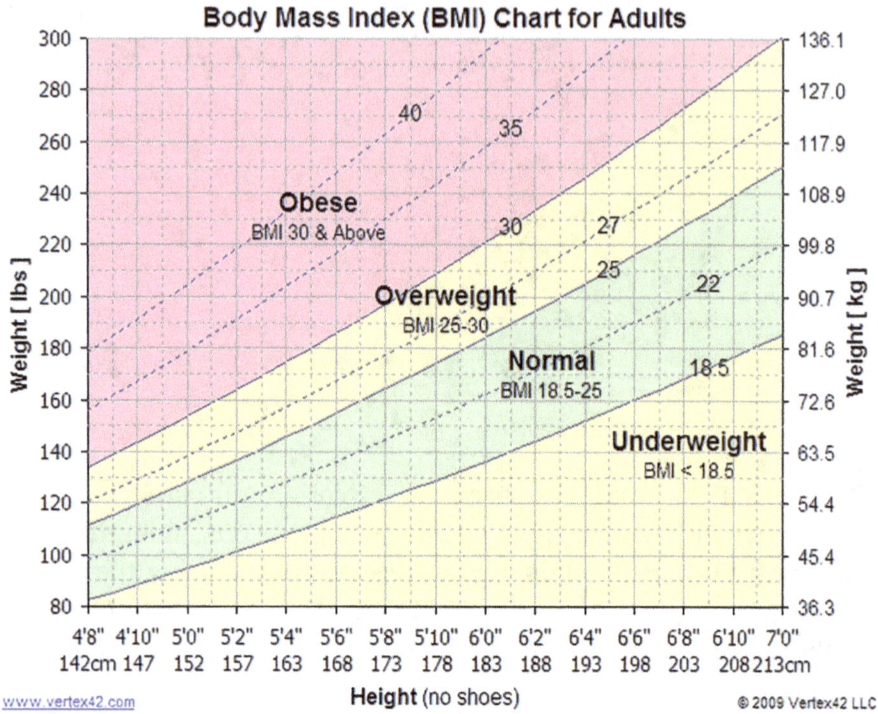

Body Mass Index Chart for Adults
Weight, Height, no shoes, Obese, Overweight, Normal, Underweight.

OBESITY

The United States of America performs 110,000 bariatric surgeries annually to deal with obesity in ever younger persons.

Personally, I have seen only two Latinos who have subjected themselves to bariatric surgery (of course there are more, but I have not seen them in my practice). The surgical specialists comment that there is an "underuse of bariatric surgery among Latinos" and ask themselves, why? What do you think? It is not for lack of obesity. Why do we resist? Perhaps we still have access to the wisdom of our ancestors. No doubt, the industry would like to see an increase of this surgery among Latinos, it is a lucrative source of income that costs on average $14,000.

Reference:
Examining the Hispanic paradox in bariatric surgery. *Surgery for Obesity and Related Diseases* 2020 Oct;16(10):1392-1400

Do not forget that the same world that promotes bariatric surgery simultaneously launches to the market 4,000 calories of food per person each day—twice as much as required by an adult—and will do anything possible to make someone buy them.

Reference:
The Incidence Of Bariatric Surgery Has Plateaued In the U.S.. *American Journal of Surgery* 2010 Sep; 200(3): 378–385.

Taste starts to develop in utero. Habits and preferences are formed very early while we are still unaware.

The young person with overweight today does not eat more rice, beans, oats, sweet potato, malanga, mango, pineapple, pumpkin, cabbage, and cauliflower, but sweetened cereals, sugary pastry, scrambled eggs, fried potatoes, chicken, and breaded meats, sausages, pizza, quesadillas, and chips. The young person today instead of using water to quench their thirst, drinks juice, soda, and flavored milk in school and at home.

According to statistics from the World Health Organization, during the past fifty years, the adult obese population in the United States has grown from less than 10% in 1974 to 40% in 2017. Mexico and other countries imitating its lifestyle are not far behind.

The chronic inflammation that we see in persons with obesity is at the core of the non-communicable diseases, such as atherosclerosis, diabetes, cancer, painful and incapacitating arthritis, and difficult gastrointestinal and neurodegenerative conditions.

Malnourishment, whether it is from lack of food or excess calories, places us at risk of dying from other causes, as we witnessed during the COVID pandemic.

> Reference:
> Once-Weekly Semaglutide in Adolescents with Obesity. *New England Journal of Medicine* 2022 Dec 15;387(24):2245-2257

The semaglutides are drugs that work only while you take them. When you stop the drug, whether due to its prohibitive cost of $1,000 a month or because of a health side effect as a direct result of them, you will recover the appetite and the lost pounds. In the best of circumstances, this represents just wasted money, but you have a right to know that the FDA has issued a Black Box Warning for this and other products in its class. Such a warning is the most grave one that could be adjudicated to any drug.

In the year 2024, the use of semaglutides, a diabetes drug that also suppresses appetite, was suggested for children with obesity.

When I ask my patients when was the last time they ate broccoli or spinach, they answer instead that they eat chicken and eggs every day, as if that was better. For more than a century and thanks to undeserved subsidies from the government granted to the cattle industry, products derived from animals (meat, dairy, yogurt, eggs, cheese) have been made available to everyone everyday and are comparatively cheaper than some produce. The success has been such that we complain when the price goes up by a few cents.

> Reference:
> USDA livestock subsidies top $59 billion. *Report of the Environmental Working Group* 2023

When we stop growing up, we should not continue growing to the sides. According to adult references, if your height is 6 feet (1.83 meters) it may be considered normal to carry 200 pounds (90kg). However, I see people not quite 5 feet (1.52 meters) weighing that and more. With the possible exception of bodybuilders who have developed an exaggerated muscle mass working them hours and hours at the gym, when a regular person reaches that weight, we are talking of stored fat.

Obesity and diabetes are not the result of the carbohydrates in grains, beans, and fruits that turn to energy, but the result of meat, cheese, eggs, oil, fried food, and factory items. What allowed obesity and diabetes to become epidemic is that we ignored the problem for far too long, and now we seem to have lost grasp of it. To solve it, we will have to invest time and money to educate people about their food, nutrients, and how the body uses them. Me first.

The house of cards has grown too big, pays many salaries, sells millions of gadgets, dupes you into continuous measuring of blood sugar, continuing education credits to physicians, and even insulin to people whose pancreas already produces it for free. It is surprising, but not unexpected, that there are fights brewing between the companies that make appetite suppressing drugs and those who want you to buy their 4,000 calories a day. Both are fighting over you. Do you feel important or manipulated?

What about water? It is essential for life, but before imitating the neighbor and go buying bottled water uninformed, check if the water coming out of your faucet is contaminated or clean. Have it analyzed; very likely this service is provided for free in your region and you would receive a written report, so you can make an informed decision.

Don't be ashamed of not knowing; that has an easy remedy. We should feel responsible for arming ourselves with the truth. That way we can make good decisions for a healthy future. Intellectual laziness, that is something else altogether.

Let us return to overweight and obesity. The following study sheds light over the consequences of using the popular ketogenic diets for losing weight. These are diets that emphasize the consumption of fats and restrict the consumption of foods rich in carbohydrates. Following such a plan, the body is forced to convert fats into ketones and even to use its muscle protein to obtain its energy, instead of glucose which is our default.

> Reference:
> Effects of a Ketogenic Diet on Body Composition in Healthy, Young, Normal-Weight Women: A Randomized Controlled Feeding Trial. *Nutrients* 2024, 16(13), 2030

This study illustrates that a strict ketogenic diet is effective for losing weight, even in normal weight women. The problem is that the subjects lost more than half of the absolute weight in lean muscle mass and not fat mass, exactly the opposite of the desired result. The authors are quick to explain that women following this diet should embark on resistance exercises to be able to retain their muscle mass. Exercise should always be part of a healthy lifestyle, but here they ask that women redouble their efforts to avoid losing muscle mass on a ketogenic diet.

The investigators present no discussion of ketogenic diets. Don't mention how these increase inflammation, or how their lack of fiber and other essential nutrients found only in whole plants is associated with increased all-cause mortality. It does ask that you correct its own self inflicted deleterious effects with more exercise. What do you think of this study?

Anyone can imagine unhealthy ways for losing weight. It is unnecessary to realize studies about such methods, when we already know effective and delicious eating patterns that help us maintain healthy weight and that as dividends grant us years free of disabilities. Why are we so reluctant to accept the truth?

In this historic moment, we are living a paradox where lack of nourishment coexists with overnutrition. When there is not enough rice, beans, fruits, and vegetables, people become undernourished from lack of calories and they look skinny; but with the constant access to pizza, chicken, eggs, cheese, fritters, and pastry, the modern problem is overnutrition due to excess calories and people look plump or stocky.

For centuries, humanity has suffered lack of reliable sources of food. We roamed hungry looking for things to eat. We collected fruits, roots, and tubers that could be turned to sugar for immediate energy. We searched for nuts, seeds, and anything like an avocado for the pleasant slippery sensation in the mouth, fat that the body could store and turn to precious energy later. We are the descendants of those who survived and grew old enough to reproduce. Equipped for dealing with short supply we never had to develop protections against excess.

Historically, we are living a novelty. Modern diseases, such as diabetes, atherosclerosis, and many cancers, show up in places of abundance, not those of scarcity. We walk a very changed world while still using the same wiring we had in the Stone Age. As a result, we are defenseless when we park ourselves before a table of deli meats, cheeses, fritters, pastries, and sweets. Remember that the food industry launches to market double the calories that we need, just for you.

NATURAL SCARCITY AND FABRICATED SCARCITY

Air, sun, and rain are gifts from nature and we have withstood a great many cycles of natural abundance and scarcity. During periods of drought when winds flog the soil without mercy or when it rains and rains and we don't see the sun for weeks, we see it reflected in the harvest. Ahead I want to call attention to something different, manufactured scarcity.

When we allow industries to discharge their waste in lakes, rivers, and oceans, we create an artificial scarcity of clean water. Commercial businesses take advantage of this opportunity to steal water that flows freely from crystal clear springs, bottle it, assign it a price, and turn it into merchandise that which nature has gifted us.

When we allow industries to discharge toxic contaminants to the atmosphere, we create an artificial scarcity of pure air. Homes located where you can breathe uncontaminated air are rented or sold at higher price, out of reach for many.

When we allow industries to discharge toxic chemicals onto the soil, we create areas of contaminated terrain. In order to certify that a piece of land is free of heavy metals and other poisons or that any produce grown in that plot is organic, it has to go through costly tests and treatments financially out of reach for many farmers.

> Reference:
> *The Gift: How the Creative Spirit Transforms the World.* —Lewis Hyde

How do we respond to these situations? Do we protect our own by filling the shopping cart with bottled water? Do we look the other way, because we are not the ones paying? Do we buy inexpensive subsidized food? Do we take advantage of the foreclosed home and auctioned property? Or do we insist that everyone follows the laws and advocate for the truth.

"All for one and one for all."—motto of *The Three Musketeers* by Alexandre Dumas

These situations of fabricated scarcity benefit an economy whose growth and success depend on selling products, even when they were stolen from nature. We have allowed some to pollute Mother Nature's gifts to then watch them sell the remaining unspoiled ones as if they belonged to them. Buyers and sellers, one way or another, we are all part of the system.

Scarcity in the midst of abundance. On this planet, we already grow enough food to feed the entire world population. However, instead of feeding the grain to people, we feed it to animals that must turn it to flesh, before they are killed and consumed.

More than 47 million people live hungry in the United States and in the year 2023, we left 783 million people around the world without food. Did you know that the distribution of arable land on the planet favors plants to feed cattle rather than people? In other words, the global hunger crisis is the direct result of the diet based on meat insisted upon by the western population.

Of the planetary surface that we can use, 50% is devoted to agriculture; 77% of that agricultural land is dedicated to cattle. The meat and dairy which results from that arrangement contribute only 18% of the global calories: a very poor return of a lopsided investment. All to satisfy the taste buds of those who insist on chewing meat, drinking milk, and frying eggs.

Bringing up cattle requires an exorbitant amount of land. This includes the space for each animal, crops to feed them and water to give them drink, not to mention the depletion and contamination of our treasured reserves of fresh water. A royal waste of land and resources!

Let us better understand this senseless distribution: one free roaming cow requires 2 acres of land and will render 400 pounds (181kg) of meat. However, in 2 acres of land, we could grow 40,000 pounds (18,000kg) of plant food. Project Raíces Global in the Dominican Republic of the Center for Nutrition Studies T. Colin Campbell, is an example of this reality.

In summary, 82% of the global nutrition comes from 12% of land that is dedicated to plants; plants that do not pass through the stomach of an animal to become meat.

We already have the resources to feed the entire world population with grains, beans, fruits, mushrooms, and vegetables. We have enough plus money and personnel left over to care for those who, by no fault of

their own, get sick with conditions that require special care. Imagine an economy centered on health instead of disease. As responsible residents of this planet Earth, we should subscribe to plant-based nutrition.

DIABETES

The pancreas of someone with type 1 diabetes does not produce insulin. Whereas, the pancreas of someone with type 2 diabetes (which is 95% of the cases of diabetes) produces enough insulin throughout most of the disease's natural course.

The pancreas of a healthy person releases between 20 and 40 units of insulin each day (an average of 27 units). Nonetheless, there are too many people with type 2 diabetes who even injecting 50 units or more of insulin each day cannot achieve normal levels of blood glucose. Obviously, the failure is not a lack of insulin.

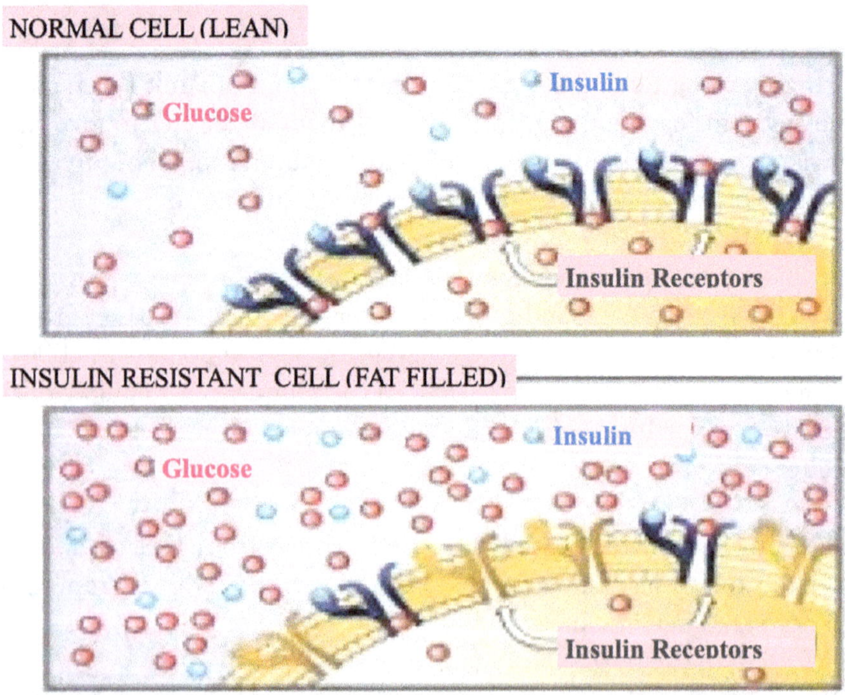

Insulin resistance starts with the accumulation of saturated fat inside the cells, which then incapacitate their insulin receptors and don't allow the glucose to enter.

We have learned that the principal problem in type 2 diabetes is insulin resistance. We know that when the cells of liver, muscle, and other parts of the body fill with fat, they inactivate their insulin receptors and don't let the glucose come in, which leaves it to accumulate in the blood. It is proven that if we lower the consumption of fat to less than thirty grams per day, we can reverse insulin resistance and let the glucose inside the cells where it will be turned into energy. This process of reversal with diet has been clinically demonstrated a great many times.

Of all patients who develop type 2 diabetes, it is difficult to know how many of them produce low enough insulin levels to require supplemental insulin. We do know that the pancreatic cells that are exposed to toxic high levels of sugar, eventually stop producing insulin. An easy to obtain C-peptide blood test lets us see the amount of insulin produced by the pancreas, you just need someone to help you interpret the results.

When we did not know how a person developed type 2 diabetes or did not know insulin resistance, the management of the disease consisted in limiting the consumption of carbohydrate rich foods, to reduce the glucose in the blood. Now that science has explained the mechanisms of diabetes, we should not insist on this antiquated management, instead we can reeducate ourselves with the truth.

REVERSE INSULIN RESISTANCE

To reverse insulin resistance it is necessary that we consume less than 30 grams of fat a day. This table demonstrates that it is much more difficult to stay below the 30 grams of fat if you consume oils, eat too many nuts and avocados, and depend on animal products.

1 Tbsp any oil: 14 g of fat
3 ounces of salmon: 11 g
3 ounces of chicken: 13 g-17 g
1 ounce of nuts: 14 g-20 g
2 eggs: 10 g
1 avocado: 30 g
1 ounce of cheddar cheese: 9 g
1 ounce of mushrooms: 0.3 g

1 banana: 0.6 g
1 cup of kale: 0.5 g
1 cup of oats: 2.3 g
1 cup of lentils: 0.4 g
1 cup of cherries: 1.4 g

In clinical studies it has been demonstrated that any person can develop temporary insulin resistance. For example, we see that after a meal rich in animal fat, fritters, cheese, or pastry, the cells fill with saturated fat and reduce the activity of their insulin receptors for several hours. If the next meal is low in fat, the cells reactivate their receptors and recover insulin sensitivity. That said, if the person insists on filling the cells with fat one meal after the next giving the cells no chance to recover, sugar accumulates in the blood more and more, and what was a temporary state progresses to prediabetes and diabetes.

DIABETES AND INSULIN RESISTANCE

Even though we are drowning in diabetes, few physicians know that in 1932, Dr Mordecai Rabinowitch in Canada reversed type 2 diabetes in a series of patients using a diet of rice and fruits. In 1950 Dr Walter Kempner in Duke University did the same thing, his diet of rice and fruits not only reversed diabetes in his subjects, but also high blood pressure and cholesterol.

It should be left clear that these are not diets to which we should aspire. They are presented to demonstrate that a diet of carbohydrate rich foods like white rice and fruits without added oil is capable of reversing insulin resistance. Using modern investigative tools, we have been able to uncover the mechanisms that explain the findings.

Reference:
Effects of the high carbohydrate-low calorie diet upon carbohydrate tolerance in diabetes mellitus; Rabinowitch IM. *Canadian Medical Association Journal* 1935 Aug 33(2): 136–144.

I repeat, the problem of insulin resistance, prediabetes, and diabetes rests on the accumulation of fat in cells (especially saturated fat) obtained from meats, milk, cheese, eggs, fritters, pastry, and ultra processed factory foods on top of the fat full cells in a body with overweight or obesity.

With the following table, you can see that foods rich in fat tend to be low in carbohydrates and that conversely, those rich in carbohydrates tend to be low in fat. Except for nuts, seeds, soy, and avocados, the only plants with more than 70% of their calories come from fats). plants are rich in carbohydrates and low in fat. As you learn more about these patterns, you will dominate the material and your practical knowledge will increase.

INGREDIENT	TOTAL FAT	CARBOHYDRATES
1 Tbsp olive oil	14 g	0 g
6 black olives	6 g	5 g
3 ounces salmon	11 g	0 g
3 ounces chicken	13 g-17 g	0 g
1 ounce nuts	14g-20 g	5 g
2 eggs	10 g	2 g
1 avocado	30 g	6 g
1 ounce cheddar cheese	9 g	0 g
4 ounces firm tofu	10 g	5 g
1 ounce mushrooms	0.3 g	21 g
1 banana	0.6 g	28 g
1 cup kale	0.5 g	7 g
1 cup oats	2.3 g	25 g
1 cup lentils	0.4 g	20 g
1 cup cherries	1.4 g	25 g

It has become fashionable to reduce foods rich in carbohydrates in order to control blood sugar levels; these diets have gained popularity among people who want to continue eating meat, cheese, eggs, and fritters. Don't bite the hook.

The ketogenic diet, low in carbohydrates and high in fat, has stuck because it includes oils, meat, and eggs—and people like that. Since there are few carbohydrates turning into glucose, the blood sugar levels will be lower, but the person will still be insulin resistant. This reverses nothing.

Absent carbohydrates, the body needs to find another way to generate energy and will do it by burning amino acids from proteins and ketones from fat. However, this is not your body's plan A and it is also not the most efficient (not to mention the risks associated with this diet).

As already mentioned, the pancreas of a normal person without diabetes produces between 20 and 40 units of insulin per day. Recently I saw a patient with type 2 diabetes who injects 60 units of supplemental insulin daily and still cannot achieve blood sugar levels below 280mg/dL (normal levels should be around 100mg/dL)—he follows a ketogenic diet. His fat filled cells are resistant to all insulin, the insulin made by his pancreas and the one he injects. To achieve a normal level of blood glucose of 100mg/dL, a person living with insulin resistance on this diet may not eat mango, banana, oats, sweet potato, potato, pumpkin, rice, or beans; a situation which he calls an unsustainable torture. And yet, he has the option of reducing his insulin resistance by eating carbohydrate rich foods that are naturally low in fat (mango, banana, oats, sweet potato, potato, pumpkin, rice, and beans), however since he does not want to eliminate his chicken and oil, he is stuck. What road would you choose?

In spite of this marvelous knowledge, many professionals insist in limiting the consumption of foods naturally rich in carbohydrates. One of the questions in my most recent family medicine board recertification exam 2025 had as the "correct" answer for the treatment of type 2 diabetes, a diet low in carbohydrates! We have returned to the times when we treated diabetes with ignorance.

HYPERTENSION AND THE KIDNEYS

The force exerted by the blood against the arterial walls is defined as arterial blood pressure. That force increases during the contraction of the heart (systole) and returns to base level when the heart relaxes (diastole). The arterial blood pressure is reported in millimeters of mercury, for example 100/60mmHg.

Some theorize that for "inexplicable" reasons the pressure inside the arteries rises and turns them narrow and rigid. This theory presents the body as something defective that we must fix with the help of our superhero, the pharmaceutical industry. We would have to ignore everything we know about the vascular system to say that we develop elevated pressure inexplicably.

Instead, the theory with scientific evidence sustains that when the endothelium of the arteries is damaged, it cannot function properly causing the arteries to become rigid and narrow. Let's see, how is the endothelium damaged and what causes this damage? We are guided by curiosity.

It is observed that when the endothelial cells are exposed to excess salt, alcohol, or saturated fat, they cannot produce nitric oxide, a gas that among its other functions relaxes the arteries musculature. Ah! It has also been observed that the accumulation of oxidized cholesterol underneath the lining of blood vessels (endothelium) narrows the lumen of the arteries and makes them more rigid. When it is we who have created the conditions for hypertension, we can also reverse them. Who is the superhero now?

We have turned loose on the environment chemical substances called endocrine disruptors in the form of herbicides and plastics that when filtered through the kidneys damage the nephron (functional kidney unit). When we add these substances to a sustained elevated pressure, we further burden these organs and damage them even more. Once the kidneys stop filtering toxins in the blood, the person depends on a dialysis machine to live.

We have good information with which to ameliorate or buffer many risks. In fact, conventional medicine recommends eating low in saturated fat and animal protein to reduce the workload of the kidneys. This very simple formula helps not only the kidneys, but every organ in the body. Why is it so difficult for us to implement? Do you have any explanation?

Reference:
The Relationship between Typical Environmental Endocrine Disruptors and Kidney Disease. *Toxics* 2023 Jan; 11(1): 32.

CORONARY DISEASE AND CHOLESTEROL

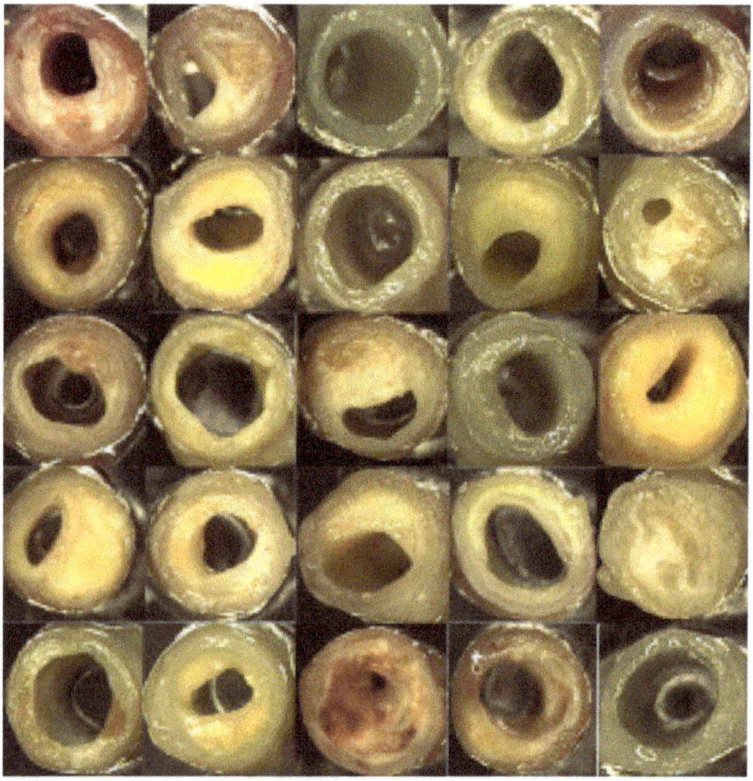

These photos of arteries with cholesterol plaques show the limited circulation they allow through their interior.

In 1968, I was a medical student in Puerto Rico and the "normal" cholesterol of patients in the coronary unit of the Municipal Hospital in San Juan was 300mg/dL. In other words, such levels were the norm. According to peer reviewed scientific publications today, we can avoid a coronary infarct by keeping cholesterol levels under 150mg/dL. This is easily achieved on a low-fat plant diet, yet much of what we have discovered is kept in the dark.

Reference:
Prevent and Reverse Heart Disease. —Caldwell Esselstyn, MD 2008

In 1976, Nathan Pritikin, who reverted his coronary disease on a plant diet and exercise, published in *The Journal of Applied Nutrition* a valuable summary titled "High Carbohydrate Diets: Maligned and Misunderstood."

We should learn about the Trimethylamine (TMAO) molecule responsible for who knows how many heart attacks. Bacteria in our intestine digest the choline and L-carnitine in red meat, fish, chicken, and eggs and produce trimethylamine (TMA). When oxidized in the liver and converted to TMAO, it circulates causing inflammation in the endothelium of every artery in the body.

During my first year of residency, I used to moonlight in the emergency room of a community hospital in Philadelphia to supplement my income. Since I had no family nearby, I offered to cover holidays. On those nights, we would receive people dressed in their gala outfits directly from their gastronomic celebrations, in a florid heart attack or plain heart failure. Without a doubt they came with elevated levels of TMAO, which we still didn't know since its implications in cardiac events was not described until 2011.

Reference:
Trimethylamine N-Oxide Generated by the Gut Microbiota Is Associated with Vascular Inflammation: New Insights into Atherosclerosis. *Mediators of Inflammation* 2020

Not long ago, we considered it essential for cardiovascular health to have high levels of HDL cholesterol. This high-density lipoprotein molecule removes cholesterol from the arterial plaques and transports it to the liver, who then discharges it onto the bile. From there cholesterol passes to the intestine and is evacuated in the stools.

The number level of HDL was considered so important that anyone with low levels was considered to be at risk of dying from a heart attack.

I remember the parents of a friend of my son's in panic during a college picnic, because the young man had low cholesterol, but an HDL of 29mg/dL and his doctors feared he could suffer an infarct at any time so were looking desperately for a way to increase the levels of HDL to at least 35. If you want an example of reductionism, here is one.

Today we know that what matters is not the concentration of HDL, but its ability to remove cholesterol from the arterial plaques, or its efflux capacity. Once we understand this concept, it would not surprise us that someone with low cholesterol levels would also have low levels of HDL without increasing his risk for a heart attack; because if they are not accumulating cholesterol, why would they need high levels of HDL, whose function is to remove it from the plaques? The body is economical, if it does not need something, it will stop producing it to save energy and resources. Similar to the reduction in lactase when we are weaned from breastmilk.

Reference:
Changes in Lipid and Lipoprotein Levels and Body Weight in Tarahumara Indians after Consumption of an Affluent Diet. *New England Journal of Medicine* 1991;325:1704-17

Therein lies a valuable lesson. Do not let your curiosity rest. Scientific discoveries are points of departure; rarely are they the final point.

HYPERTENSION AND THE HEART

Different from most muscles which are solid like the biceps, calf, or tongue, the heart is a pump with four cavities (two auricles and two ventricles) complete with plumbing through which it fills and empties itself.

When the heart dilates, the veins fill the auricles with blood which immediately passes to the ventricles. When the heart contracts, the blood shoots out of the ventricles through the aorta and the pulmonary artery. The heart pushes the blood through these two principal trunks; the blood does not move unless the heart can overcome the resistance of the entire column of blood in order to advance the circulation.

If we make the heart work against high resistance (as in elevated blood pressure), it's like having it live at a gym 24/7. Its muscle fibers get thicker, they grow in number, and the muscle gets bigger. When the heart muscle gets larger, its cavities get smaller, and can pump less blood with each heartbeat.

Sustained high pressure causes double damage to the heart: the ventricles fill with less amount of blood and the heart has to pump faster to move the same volume.

The arteries of someone on a diet rich in saturated fats do not produce enough nitric oxide to keep them elastic. At the same time, the inside of the arteries keeps getting narrower with cholesterol plaques that obstruct the blood transit even more.

Every cell needs a certain amount of cholesterol, but when they have enough the cells inactivate their membrane receptors and leave it circulating in the blood, where it is exposed to oxidation. The macrophages that live beneath the arterial endothelium gobble it up and form deposits in its walls—which are the beginning of cholesterol plaques. The endothelium over that stripe of cholesterol stretches very thin and fragile. If the plaque survives this early stage, it gets plastered with scar connective tissue and calcium making it durable. On the other hand, if the force of the blood flow breaks the young plaque and spills its inflammatory content onto the blood, it provokes a sudden clot and the event we know as a heart attack or infarct.

That clot corks the inside of said artery leaving without oxygen and nutrients all the tissue downriver that depended on it. As a result and without any alternate route to pick up the slack, that portion of the heart dies. In the United States there are around 700,000 lethal heart attacks a year in which the artery that gets occluded is too big and the person does not survive. In Mexico there were 226,000 deaths from heart disease in 2021. The sad thing is that more than 90% of infarcts can be avoided following a lifestyle free of tobacco and a diet based on plants low in fat.

If the obstruction of the coronaries occurs very slowly taking decades, this could provide the opportunity for the formation of alternate or collateral coronaries that would supply circulation to the areas more compromised due to this hardening of the arteries also known as atherosclerosis.

The blood circulation of the heart muscle is very simple and different from the brain which is designed in advance with collateral circulation that allows its tissues to be irrigated through various routes. The circle of Willis in the brain shows a unique anatomical vascular difference that helps us explain why coronary infarcts occur at much younger age than cerebral infarcts, as we will see later.

In the 1990s, the National Institutes of Health (NIH) led us to believe that modifications to the American diet could help reduce hypertension, a significant risk in heart disease. It is a shame that the recommended dietary approaches to stop hypertension (DASH) did not keep its promise.

Originally, the DASH focused on reducing sodium (Na) to 1,500mg a day and emphasized eating minimally processed fruits and vegetables each day (instead of refined carbohydrates). Note that one pound of American cheese can pack up to 7,500mg of sodium. But as a result of the objections from the industry, the National Institute of Heart, Lungs, and Blood backpedaled and the diet was launched to the public with the unmistakable fingerprint of the cattle and dairy industry: the recommendation to consume no more than 2,300mg of sodium, lean meats, and low-fat dairy products.

Bending knee to the industry, we ceded the vanguard. When the recommendations were relaxed, the excuse offered was not wanting to upset the public, but who did they really keep happy?

All in all, the touted diet against hypertension or DASH recommends a calorie distribution almost the same as the existing standard United States diet.

DASH: 50%-55% carbohydrates, 15%-20% protein and 30% fats versus:

Standard Diet: 50%-60% of calories from CHO, 15%-20% protein and 30%-35% fats.

This diet attracts educated followers because it is similar to the popular Mediterranean diets that people associate with health since they include olive oil. (If the Mediterranean diet has benefits, it is in spite of olive oil and not because of it.)

Too bad that after so much hullabaloo, the DASH diet does not even comply with the requirements to reverse insulin resistance and other chronic conditions.

HYPERTENSION AND ERECTILE DYSFUNCTION

Erectile dysfunction was previously known as impotence, which meant, inability to consummate the sexual act due to any number of reasons: anxiety, depression, stress, medication side effects, diabetes, or atherosclerotic disease. The public, however, interprets impotence as lack of masculinity, therefore, the euphemism—erectile dysfunction—had immediate acceptance. This new term also paved the way for the pharmaceutical industry to do public propaganda of the pills sildenafil and tadalafil (nitric oxide donors) without having to deal with the prejudice. Erectile dysfunction is a recognized marker of heart disease. It is understood that the younger the man is when presenting with erectile dysfunction, the more serious and aggressive is his atherosclerotic condition.

Reference:
Erectile dysfunction and cardiovascular disease. *Arab Journal of Urology* 2013 May 3;11(3):212–216

HYPERTENSION AND THE BRAIN

To think of the body as a collection of organs that function separate from each other is to ignore that the body is like an orchestra and much more than the sum of its parts. In an integrated organism, what affects one part affects all the others. What benefits one, benefits the others.

The brain is the mother board. Do you remember the alternate routes the heart can slowly create to deliver oxygen and nutrients to compromised areas? Well, the brain was born already with such a system; it comes equipped with a web of alternate routes. The carotid arteries, the cerebral arteries, and the communicating arteries connect all other blood routes in the brain, so that if access to one area is blocked, it can be reached via another.

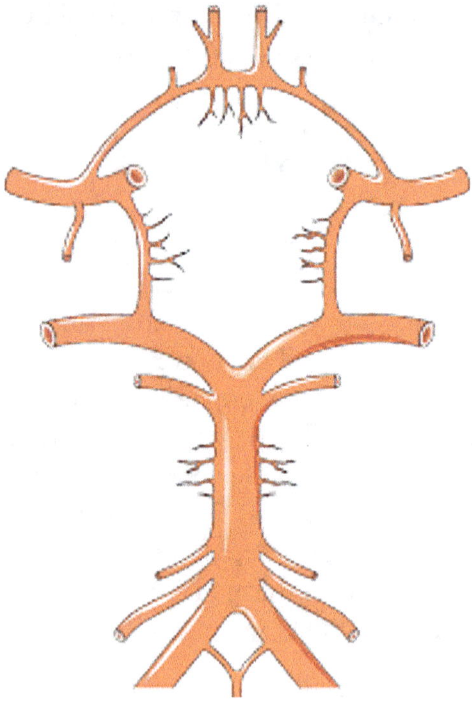

The cerebral circulation through the Circle of Willis offers a great advantage over the coronary circulation of the heart.

When we abuse this marvelous design with the western diet, there are consequences. What damages the coronary arteries is also damaging the cerebral arteries—the cholesterol plaques, the stiffening of the arteries, and the chronic inflammation are compromising the oxygenation and nutrition of brain cells. Even when these do not provoke a massive cerebral thrombosis, little by little they cause microinfarcts, microaneurysms, degeneration of the cerebral matter, and in time dementia.

During my medical training, we spoke of cerebral microinfarcts, micro aneurysms, and dementia as if they were separate entities. Those were other times and we were ignorant. What is inconceivable is that we are still speaking that way. Atherosclerosis is a vascular disease, not simply cardiac, renal, erectile, vertebral, peripheric, or just cerebral. We own the whole package, we are responsible for the human body as a whole.

When the brain presents with dementia, we can infer that its owner has subjected it to excess fat and protein, simple sugars, chronic inflammation, and oxidative stress for a period of sixty to eighty years, and that its cells could not dispose of the waste efficiently. According to doctors Dean and Ayesha Sherzai, over 90% of Alzheimer is associated with dysregulation of glucose, dysregulation of fats, chronic inflammation and oxidation. In other words, brain disease is happening alongside and for the same reasons as coronary disease.

Reference:
Preventing Alzheimer's: Our Most Urgent Health Care Priority Dean Sherzai, MD, MPH, PhD and Ayesha Sherzai MD. *American Journal of Lifestyle Medicine* 2019 Sep-Oct; 13(5): 451–461.

MENOPAUSE IS NOT A DISEASE

Just as every girl has a menarche (first menstrual period), menopause is the time in a woman's life characterized by profound changes in the levels of reproductive hormones. It is not an abnormality that should be medicated just because, even when the changes bring physiologic adjustments that we experience with more or less elegance. Some women hardly notice it, while others report a storm with alterations in thermoregulation, hot flashes, sweats, irritability, and some degree of vaginal atrophy. Each woman will report particular symptoms that can last days, weeks, or years. The healthier her diet and weight have been before she entered this period, the more integrated and less disruptive might be the course and the less marked might be the changes relative to fat and muscle in the body.

We should be able to alleviate some symptoms and smooth the way with a whole foods plant diet, increasing the ingestion of soy, consuming less fat, returning to a normal weight, exercising regularly, and paying attention to the state of our mental and emotional health.

The use of topical hormones and even the replacement of hormones is within reach and must be individualized for each woman, taking into account risks and benefits. There is not one single way to live the menopause.

In addition to internal hormone changes, menopause has a cultural dimension. The woman in our society is forced to maintain an artificial appearance of eternal youth. If she does not have support to navigate this time, it will be all that more difficult. Many studies have brought to light the fact that aging women who live in cultures where they are recognized, respected, and celebrated in all their glory, have a menopause experience very different than women who are required to dye their grey hair, have plastic surgery, and live always appearing to be young; or worse where she is called old, a witch, and a harpy. The older woman in literature is held in many ways: from a despised character to a venerated presence.

In 1963, the Brooklyn gynecologist Robert A. Wilson and his wife Thelma published an article in the *Journal of the American Geriatrics Society*, where they argued that menopause robs a woman of her femininity and ruins her quality of life.

Such interpretation of what it means to enter this natural period in a woman's life was pushed forward in 1966, when the same Robert Wilson published a best seller book titled *Feminine Forever* where he maintains menopause is a disease and should be treated with estrogen to counteract the inevitable "wearing down" of the woman. Do you subscribe to this notion?

OSTEOPOROSIS AND BONE FRAGILITY

One fact that merits attention is not the mere porosity of bones (osteoporosis), but the fragility of such bones. This refers to the risk of fractures due to very slight trauma. Bones are complex living tissue in constant remodeling whose resistance to fractures is assured with way more than a mineral, for example with an alkaline diet (as you will see) and regular exercise in the sun.

In the 1830s, the only way to diagnose bone porosity was in pathology after the person died with a fragility fracture. When technology allowed us to see porosity before a fracture occurred, the condition of osteoporosis became a diagnosis and a disease that required treatment. In 1987, with the use of the instrument we know as DEXA scans, over 60% of women over 40 years were diagnosed overnight with osteopenia and osteoporosis. The medical complex sowed fear in all women, which added to the growing sentiment of having to stay eternally feminine and the flourishing industry of hormone replacement was born. What a success.

Fear contributes to health erosion. When Carina was young, her diet was very processed, she was always active, walking natural trails, jogging, swimming, and cycling. Given her lifestyle, she has fallen various times, but never broke a bone. During a routine densitometry five years ago, she was diagnosed with osteoporosis. Without any discussion, she was prescribed injected osteoporosis medication. Ignorant of who owns this skeleton, the specialist painted her a picture of bones that would crumble if she does not follow the treatment. As a result the once confident athlete, does not take a step without fear of breaking.

Medicine offers too many examples of this reductionistic way of interacting with patients. We are trained with a hammer to only hit a nail.

In fact, we want to avoid bone fractures, not simply bone porosity. The importance of osteoporosis not withstanding, the broken bone of someone who lives an active and healthy life will heal. The cells responsible for depositing and maintaining calcium in the bones respond to the quality of our diet and our physical activity. Bones are living organs demonstrated to be in constant dynamic remodeling.

Regarding the diet, the minerals phosphorus and potassium favor the retention of calcium in the bones while animal protein requires calcium from the bones to neutralize the acid pH. The loss of calcium in the urine is corrected by absorbing more calcium during the next meal. However, if the dietary pattern favors a loss of calcium, of course the bones will become more porous. The idea is to prevent, rather than chase the loss.

The cells of bones that are exercised against the forces of gravity tend to retain calcium. Any exercise helps preserve muscle strength, which is essential to maintain balance. If you don't fall, you are less likely to break a bone.

Reference:
Building Bone Vitality: A Revolutionary Diet Plan to Prevent Bone Loss and Reverse Osteoporosis by Amy Lanou, PhD and Michael Castleman

ARTHRITIS

Any pain due to inflammation in a joint (the place where two bones meet) is called arthritis. The most common of them is osteoarthritis or degenerative arthritis, whose pathologic characteristic is worn cartilage. The joints most commonly affected are hands and knees of people with overweight and obesity, in whom even the adipose tissue of the bone marrow becomes inflamed. The pathologic changes of osteoarthritis cause pain, stiffness, and restrict movement. You may imagine the benefits of an anti-inflammatory plant-based diet, which will also help with weight loss, as indicated in the following studies.

References:
Whole-Foods, Plant-Based Diet Alleviates the Symptoms of Osteoarthritis. *Arthritis* 2015 Feb 28;2015

Effect of a multidisciplinary lifestyle intervention on body composition in people with osteoarthritis: Secondary analysis of the "Plants for Joints" randomized controlled trial. *Osteoarthritis and Cartilage Open* 2024;6(4)

SLEEP APNEA

An estimated 26% of adults in the United States suffer obstructive sleep apnea. Together with obesity, this condition has been on the rise in younger and younger patients. It is associated with greater risk for hypertension, cardiac disease, type 2 diabetes, cerebral strokes, and depression.

Although not proven to be beneficial, the treatment most commonly prescribed is for the person to sleep connected to a machine that exerts continuous positive pressure (CPAP) inside the airways. Without health insurance, the machine costs between $500 and $1,000 plus another $200 for the tubing and mask. The medical complex avoids dealing with the two most important risk factors: diet and obesity. Welcome to medicine by compartments.

> Reference:
> Can Continuous Positive Airway Pressure Reduce the Risk of Stroke in Obstructive Sleep Apnea Patients? A Systematic Review and Meta-Analysis. *Public Library of Science PLoS One* 2016 Jan 5;11(1)

CANCER

We could inherit a genetic mutation that increases our susceptibility to suffer from a particular cancer, but cancer itself is not hereditary. To develop a cancer, it is first necessary to suffer damage in a chromosome and such damage can be caused by radiation, a drug, a carcinogen in the diet or the environment, in essence by any one of thousands of possible agents. In the passage ahead, I will try to summarize what we understand about cancer.

We have genes that supervise and control the multiplication or proliferation of our cells. The p52 and p53 are two genes important in this task. Every time they detect a cell with damage in a chromosome trying to divide, they destroy it in the act. This constant vigilance prevents a defect from passing on to the next generation of cells and tries to assure that only healthy cells can successfully reproduce. The to and fro between this promotion and repression can last decades. It is important to remember that our nutrition exerts a powerful influence over this process.

For example, eating foods with animal protein rich in growth hormone (meat and milk) promotes the multiplication of cells. Add to it the excess of free radicals (resulting from processed food) and we impede the function of our protective mechanisms. Living in a state of chronic generalized inflammation, we create the conditions for our system of supervision and vigilance to fail. On the other hand, we could support our systems of control and equilibrium by reducing the production of free radicals, eliminating the stimulus for cells to multiply through growth hormone, augmenting our antioxidant capacity, and fortifying our intestinal flora through a diet of an abundant variety of whole plants. The evidence and mechanisms of these phenomena came out of the investigation of T. Colin Campbell and his colleagues in The China Study.

Following a diet of whole plants where we avoid animal products and processed foods is associated with the repression of potentially cancerous cells. Various clinical studies have shed light on the effects of a diet based on whole plants, not only in its preventive role but in declared cancer as well, like the studies that have come out of the investigative center of Rochester University and those in the body of work of Dean Ornish. We need many more published studies to make it more likely that practicing physicians find and recognize this information, which to date is absent from school curricula or the required continuing medical education.

References:
Regulation of p53 tumour suppressor target gene expression by the p52 NF-κB subunit). *The European Molecular Biology Organization EMBO Journal* (2006) 25: 4820 – 4832

A whole-food, plant-based randomized controlled trial in metastatic breast cancer: weight, cardiometabolic, and hormonal outcomes. *Breast Cancer Research and Treatment* (2024) 205:257–266

AUTOIMMUNE DISEASES

In autoimmune diseases, the same system that protects us from microbial invasions and pathogens unleashes an auto destructive reaction. Chronic inflammation, along with genetic and environmental factors such as the diet, play a crucial role in these diseases. An anti-inflammatory diet based on whole plants is used successfully to calm the destruction perpetrated by antibodies in autoimmune diseases.

Type 1 diabetes is an autoimmune disease. Here the pancreatic cells that produce insulin are the target of destruction. It has been found that cow's milk consumption at an early age is associated with an increased incidence of this disease.

Perhaps you know of Hashimoto's thyroiditis, in which the immune defenses are directed against the thyroid gland. People with overweight who follow a proinflammatory diet are most affected.

You might also have heard of psoriasis. In this case, a protein of the streptococcus bacteria which causes tonsillitis co-reacts with skin cells and sparks a chronic inflammation.

In rheumatoid arthritis, we see how the western diet of meats, cow's milk, and processed foods also increases the risks for this disease.

The destructive demyelination of the nervous system in multiple sclerosis causes progressive incapacity. Diet and the microbiota appear to play an essential role in the risk and management of this neurodegenerative disease.

> Reference:
> *Code Blue* documentary with Saray Stancic MD, author of *What's Missing from Medicine: Six Lifestyle Changes to Overcome Chronic Illness*

Just as the other autoimmune diseases, lupus is mediated through chronic inflammation. Diets rich in the saturated fat of animal products encourage a certain species of bacteria in the colon that increase inflammation and intestinal permeability.

Crohn's inflammatory disease is not classified as autoimmune, since we don't have evidence of a mechanism from the immune system attacking healthy tissue. However, as other inflammatory diseases of the intestinal tract, it is associated with the elevated consumption of saturated fat (animal protein), refined sugars, and a diet low in fiber.

References:
Dietary factors and risk of islet autoimmunity and type 1 diabetes: a systematic review and meta-analysis. *The Lancet* Volume 72,103633 October 2021

Good-bye Lupus: How a Medical Doctor Healed Herself Naturally With Supermarket Foods, Brooke Goldner, MD

The Association between Total Protein, Animal Protein, and Animal Protein Sources with Risk of Inflammatory Bowel Diseases: A Systematic Review and Meta-Analysis of Cohort Studies. *Advances in Nutrition* 2023 May 14;14(4):752–761

CHRONIC INFLAMMATION AND HEALING

Acute inflammation is one of the most dramatic responses the body ever displays. The immune system recruits the necessary help, goes to the injured area, neutralizes the damage, and begins the healing. After an injury or a brief exposure to a noxious agent, our system responds with inflammation characterized by swelling, heat, pain, and redness at the site. This response is measured, precise, and local. The situation is resolved in a matter of hours or days and can save our lives.

In contrast, if the damage persists, inflammation turns chronic and the free radicals of the immune response unleash a reaction that destroys cells of healthy tissue. Diseases such as asthma, cancer, atherosclerosis, diabetes, and also autoimmune and degenerative diseases, are associated with the persistent damage of a chronic inflammation that does not know how to stop.

A key ingredient in the chronic inflammation of these diseases seems to be the western lifestyle and its diet centered on processed foods (made with flours, added sugars and oils) and of foods of animal origin (meats, dairy, eggs).

This does not mean that a diet centered on whole plants is a cure-all, but it is a tool within everyone's reach and has only beneficial side effects. Implementing an anti-inflammatory diet based on plants does not require medical insurance, pre-authorization, or anyone's permission and it is inexpensive.

Don't delay it any longer, clean your refrigerator and pantry. Avoid falling into the habit of eating out and see how much health you can recover.

Wellbeing and health are universal themes. By observing the human body in all circumstances, we obtain valuable information that we hand from person to person and generation to generation. The contributions to modern medicine are quite recent adding themselves to an enormous accumulation of empiric knowledge. Keeping this perspective, we are enriched.

Sanatoriums became popular from the mid-19th century to the early twentieth century as places of healing. These were not hospitals, they were places where patients with chronic conditions went to be out of doors, enjoy the sun, breathe pure air, exercise, rest, and be nourished. These convalescent homes provided peace and quiet, who wouldn't like it?

Today, medicine is recognizing the need for these valuable components as part of any recovery. Surgery and pharmaceuticals cannot function alone.

The Health Center True North in California, founded in 1984 by doctors Alan Goldhamer (coauthor of *The Pleasure Trap*) and Jennifer Marano, is an exceptional place that offers prolonged water only fast under medical supervision and a diet of exclusively whole plant foods with the objective of optimizing health. Backed by published clinical studies and accepted by conventional medicine, the water only fast accelerates metabolic changes that take months when we try to make the dietary changes during our normal life routine. Although in its beginnings it was scorned, fasting has been embraced by established medicine as a valuable tool, with some modifications.

The conditions that lend themselves to this protocol are those associated with over-alimentation, such as obesity, diabetes, and many cancers. When the body is deprived of the food to which it is addicted—fat and calorie dense, high in sugar and salt—we eliminate the generous dose of dopamine that comes with it making it possible to get out of the pleasure trap in half the time. That alone offers incalculable health benefits.

The most apt candidates for this protocol are highly motivated people, not just for the fasting period, but who want to continue the new alimentation and lifestyle once they return home. The success of the program (which has seen 20,000 people) is in great measure due to the auto selection of the participants.

The benefits of this practice which has existed as part of every religion and philosophy for all time, continue to pan out. For example, according to clinical studies, in a fasting state the drugs needed to treat mortal illnesses have greater effect, cancerous cells characterized by their accelerated metabolism are more vulnerable when we deprive them of food, while normal cells seem to be protected during this metabolic rest.

The stay at a sanatorium, with or without prolonged fasting, can support the healing and recovery of which the body is capable, by removing all obstacles from the way for a period of time. True that when the person returns to his environment, he returns to the excess of products made with concentrated sugar and oil, which presents a long-term challenge. However, having recalibrated the satiety sensors during several weeks, his brain could undertake in earnest the task of creating a healthier environment and figure out how he wants to interact with the drug food and the people with whom he will associate from then on. Will is might.

LABORATORY TESTS

Blood tests are like a snapshot, they measure "something" in the blood at the time the sample was taken, be it glucose, cholesterol, sodium, hepatic enzymes, etc. The blood test reflects a moment in the exquisite metabolic equilibrium that the body regulates each step of the way.

You might ask, how can we trust laboratory tests that show just an instant if we are so dynamic? Let's see. For each test, we took many snapshots of many so-called normal and healthy people. Those results are then accepted as the normal range for the population. Those are the values we use to compare your own results.

The normal limits for some tests are very narrow, such as for blood acidity or pH; others are very wide, such as those for vitamin B_{12}; still others, such as the number of white blood cells, have no established normal levels.

In spite of the widely practiced annual exam habit, there are no established laboratory tests that a normal person should have on an annual basis. In fact, there is no consensus of whether an asymptomatic person should submit herself to an annual exam or even a regular series of tests.

Why do we use laboratory tests? The U.S. Preventive Services Task Force publishes recommendations for screening tests and qualifies them according to accepted risks in each category for age and sex. The primary screening tests are done in people who present no symptoms. An example would be the blood pressure checks offered at health fairs.

Secondary tests are used to confirm suspicions according to what is known of the person's clinical history, lifestyle, and findings on physical exam. For example, in people with overweight, one would activate tests to detect insulin resistance and perhaps markers for vascular disease.

The fracture of a bone resulting from slight trauma, would call for tests to detect osteoporosis. If blood pressure is elevated, we would test for heart disease. And so on.

If a person has a mass, a lesion, pain, is bleeding, or already has a confirmed disease, we use tertiary tests to evaluate the grade or level of the condition.

Blood tests usually confirm what we already suspect clinically, given the habits of nutrition and exercise, the toxins, and the stresses to which the person is known to be exposed. Tests results are added to symptoms, appearance. function, and physical exam, they rarely speak by themselves.

Someone on a western diet of meat, chicken, cheese, eggs, processed food, and fritters, who is overweight and experiences fatigue with exercise could have laboratory results considered normal today and tomorrow suffer a heart attack.

Do not depend on laboratory tests before deciding to improve your diet and lifestyle. A stitch in time, saves nine.

PART FOUR

General Topics of the Health Care System

Wishing health could come in a pill, we continue to poison ourselves with junk food and bad habits.

PHARMA DESERVES A SECTION

The first pharmacy in the United States opened its doors in 1885 when some wholesale businesses transitioned to the drug market. By the early twentieth century, the public already had access to all kinds of drugs for pain, inflammation, acid stomach, fatigue, constipation, insomnia, and more. Pharmacists dedicated themselves to packaging drugs for patients and the pharmaceutical turned its focus on regulation, adjustment of prices, and product innovation.

References:
Pharmaceutical Manufacturing in America: A Brief History. A Daemmrich *History of Pharmacy and Pharmaceuticals*. Vol. 59, Issue 3. 25 Jul 2017

Historical sketch of modern pharmaceutical science and technology (Part 3); From the second half of the 19th century to World War II. 1995. 30(1):1-10. Yakushigaku Zasshi

HEALTH IS FREE, DISEASE IS LUCRATIVE

Many advances in pharmaceutical and public health allow us to live freer and safer. For example, we get vaccinated against contagious diseases that in many cases would be lethal, such as polio, tetanus, influenza, COVID, hepatitis, and measles. We are able to detect public threats such as tuberculosis, Streptococcus infections, papilloma, HIV, and syphilis. We conduct disease investigations using ever better methods.

These tasks are placed in government hands, because we don't want them subject to private economic influences or to lucrative interests. But, according to stockholders of a pharmaceutical company healthy people do not use enough drugs, in other words do not generate the same degree of profits as sick people. We must continuosly advocate for preventive practices.

As citizens, we are responsible for fair regulations to be obeyed. Together we form a security web. Yet, we have permitted government representatives to be influenced by the medico-pharmaceutical complex.

The public has let go of the reins of prevention and we are sliding toward a poorer state of health. Today, over 80% of non-communicable diseases are associated with our food and lifestyle. Specialists in public health and in primary medicine expect a modest economic future. While those in super specialized medicine, the ones that manage the symptoms of preventable diseases with drugs and procedures, are generously rewarded. We exalt them, even when we pay through the nose. We keep our eyes shut as to continue graduating physicians without knowledge or expertise in nutrition or lifestyle.

I know, because I am a product of such a system. I practiced according to it for over a decade. After listening to a few reductionistic chats on medical nutrition, what did I know of the relationship between eating chicken and high cholesterol? What could I say about damage to the endothelium? Or about extra virgin olive oil and diabetes? The only language we had came from the propaganda and we used it believing it was science. The year 1987 marked a decisive turn for me, as you will see later.

DEAR COLLEAGUE

Do you ever remember studying healthy persons? I began my career with animals and a cadaver soaked in formaldehyde. From there I transitioned to hospitalized people in critical conditions. Later on, I inherited patients with chronic diseases. I don't remember having evaluated anyone healthy.

I had professors that were excellent diagnosticians, who trained us to identify diseases, but not one spoke of how diseases develop. They graduated us with a license to prescribe pharmacological agents, without a word about the enormous power that nutrition holds in its ability to protect health or to recover it.

Today, the industrial medical-pharmaceutical complex in the United States has enormous centers dedicated to the management of arteries blocked by cholesterol. Annually, it signs death certificates of more than 800,000 patients who suffered a lethal heart attack, a disease that could be 95% avoided. It employs more than 1,768,446 professionals in nursing for coronary care units. It manages around 7,500 dialysis clinics that on a daily basis filter the blood of patients whose kidneys no longer function.

This medical-pharmaceutical industry relies on 130 million clients with prediabetes and 34 million with declared diabetes. The pancreas of 95% of them produced enough insulin, but their bodies could not use it because it was blocked by fat. It is difficult to estimate how many still produce insulin. Ignorant of what caused their insulin resistance, millions inject it daily to the tune of $6,000 a year plus the cost of the strips to check the blood sugar, other oral or injectable medications, frequent medical appointments, work absences, transportation, child care... The list is long. For susceptible people, diets rich in saturated fats (with meat, chicken, milk, eggs, cheese, pastries, and fritters) result in type 2 diabetes.

The so-called western diet plus its stressful and sedentary lifestyle have resulted in obesity, heart attacks, cancers, and every day more demented people with Alzheimer's. The reductionistic medicine paradigm sees each disease as a separate entity, capable of generating income.

When we assume that each disease exists isolated from the other factors that place us at risk of other diseases, medicine has divided itself in multiple branches disconnected from the trunk. Each specialty lives in a silo without a head.

My medical school class got together in Puerto Rico for its 50th anniversary.

In 2023, my medical school class celebrated its fiftieth anniversary. Under a blue sky and caressed by the tropical breeze, just like in our beginnings, we shared news and accomplishments, got up to date with each other, remembered those no longer with us. I was sure to see the advances in medical sciences having to do with nutrition and chronic illnesses, reflected in our menu. To my surprise it consisted of fried meats and root vegetables swimming in oil, so I requested a plate based on plants. While I waited for my lunch and everyone else ate, a decorated colleague told us he has twelve stents in his coronary arteries, as narrating a marvelous technological feat. He did not seem to realize that his medical paradigm had actually failed him.

We had the chance to share from our personal and professional life as slides were projected onto a wall. Everyone wanted to be recognized and there was much about which to be proud.

I spoke of how I learned that a plant-based nutrition could prevent and reverse diseases, shared patient anecdotes, spoke of diabetes in remission, and of my work with the Latino community. I showed studies of Caldwell Esselstyn and Dean Ornish on reversal of coronary artery disease, talked about lifestyle medicine, of the Blue Zones, of The China Study and T Colin Campbell. Only a handful had heard of these investigators or their findings. Medicine keeps its members in well-guarded silos. On my way to the anniversary, I was dreaming of forming a small group to mentor young students in Puerto Rico, in appreciation for the marvelous opportunity we had to become doctors.

According to the International Federation of Diabetes, 20% of the adult population in Puerto Rico has diabetes (400,000 out of 2 million). I thought curiosity would help us forge some alliances and that we could create some educational opportunities. We have the responsibility to pay back to the school that gave us our education. If memory serves me right, tuition for our medical school at the time was about $12,000.

Once home from the trip, I tried to keep the conversation going with some colleagues. I am sad to report that nothing came of it.

It always intrigued me that so many pharmaceutical companies set shop in Puerto Rico—eleven out of the twenty principal global pharmaceuticals have headquarters in the island. Today, half of the best sold drugs in the world are made in Puerto Rico. To understand this, we need to go back to the 1970s when the federal government granted the pharmaceutical industry tax exemption on the revenue of their earnings on the island. Could there be a connection?

We must participate in the creation of a socio-economic-politic model where the government guarantees access to basic health services, education, and pensions, we cannot leave it up to private corporations.

In the year 2022, the United States wasted $4.5 trillion in the management of chronic diseases, the equivalent of $13,493 per person, we cannot say that was an investment. This represents 17% of the gross national product (GDP). To say that health is expensive is to perpetuate a lie. The stockholders of the industries that dispense insulin and viagra greatly benefit, lining their pockets with money from bypass surgeries for clogged arteries and the highest salaries in medicine for the reconstruction of joints worn due to overweight and obesity. This is not health care.

There are more than 1 million replaced knees and hips each year and around 370,000 coronary bypass surgeries. These are the procedures that generate the most money for hospitals. The cost of knee replacement in 2020 was between $30,000 and $50,000 and a bypass surgery could cost between $30,000 and $200,000. These statistics are mind blowing.

> Reference:
> Hospital and Surgeon Medicare Reimbursement Trends for Total Joint Arthroplasty (TJAs). *Arthroplasty Today* 2020 Sep; 6(3): 437–444.
>
> Centers for Medicare & Medicaid Services | CMS (.gov)

We evolved alongside the gifts of the earth. We have experience with pumpkin, rice, corn, garbanzos, papaya, apple, mushrooms, coconut, flax, and sesame seed. We have intimate knowledge of those fibrous packets filled with nutrients bathed in water.

Today, we are up to our eyeballs in meats, milk, cheese, eggs, and fritters, food that our bodies manage very poorly. For many, fast food has replaced vegetables, beans, fruits, mushrooms, raw nuts, and seeds.

Our protective feedback sensors turn useless with refined flours, extracted oils, sugar outside the cane, naked nutrients without fiber, with dehydrated preserved food that lasts months or years, food refused by insects and microbes.

The problem is not the body or the brain, they would naturally make us prefer sweets and fats. For the majority of people, the problem does not arise from a psychological trauma either. The problem is that we never had so much access to sweets and fats all the time. Nor to the novelty combinations of sugar, fat, and salt created in gastronomic laboratories.

Not everyone shows the same propensity to get addicted to these products. But the industry works day and night to make them irresistible.

You might need psychological help, but the principal problem is in an environment for which we are not prepared. Concentrate on changing your surroundings. Free yourself from processed food and you will have taken a giant step forward.

THE MANAGEMENT OF DISEASES

Medical associations sell continuing education credits (CMEs) to indoctrinate clinical providers on the management of type 2 diabetes with drugs and more drugs. At these conferences, tools to reverse diabetes using food and lifestyle are nowhere to be seen.

We have enough scientific information to render us capable of avoiding so many diseases. Sadly, it is not shared in medical schools with the students. Although this information saves lives, it is not included in the mandatory continuing education credits nor does it appear in Reference: programs such as UpToDate, that millions of professionals consult when treating their patients. Every year, 28,000 doctors in medicine and 7,000 doctors in osteopathy graduate from a total of 150 programs in the country, with hardly any practical knowledge in nutrition.

When over 80% of disease is due to diet and lifestyle, what's the use of relying on one million physicians and 149,000 doctors in osteopathy ignorant in both areas? Patients should appeal to the common sense of professionals (their physicians, nurse practitioners, and dietitians) and demand that they learn to guide them on how to follow the healthy diet that we already know prevents and reverses disease.

Your physician or health professional should provide you with references on how to lower your cholesterol or blood pressure with nutrition, but if you find yourself more educated than they, share your sources of scientific impartial information: The T Colin Campbell Center for Nutrition Studies, The Physicians Committee, the work of doctors Caldwell Esselstyn, Dean Ornish, Adriana Cortés, among others.

If you are offered printed information, verify who the authors are and do not accept leaflets posing as education for the patient generated by the Cattle/Beef industry, the Dairy Council, the Olive growers, the Egg board, or similar for profit entities. No matter if they are endorsed by the American Medical Association, the American Diabetes Association, or the American Heart Association—request and demand that the information be truthful. With the help of all, medicine will be a profession independent of corporate ties.

THE PUBLIC HAS A RIGHT TO KNOW

We have evidence that with our diet and lifestyle we could avoid heart disease, the principal cause of death, by 98%. We could also reduce the risk of the feared type 2 diabetes. By the same token, we would see a reduction in many types of cancer, autoimmune diseases, and neurodegenerative diseases. A healthy body runs a much lower risk of dying prematurely from infections, something seen clearly during the COVID pandemic.

If the professional is not curious and checks her sources, she will add to the problem repeating propaganda nonsense: that taking calcium supplements helps our bones (not so), that dairy is essential for calcium (nuh-uh), that complete protein is only found in animal tissue (a lie), that we should consume more protein (without knowing how much is enough); that we should eat meat (who said so?); that consuming soy increases the risk for breast cancer (just the opposite); that men who eat tofu develop breasts (there are more phytoestrogens in beer). When these false assertions come dressed in a white coat, they are extremely harmful, making impartial and correct public education even more difficult.

CONFLICTS OF INTEREST

The government nutrition programs are key in the propagation of disinformation. For example, the United States Department of Agriculture publishes the nutrition guide for the country, but is also responsible for protecting the cattle industry with subsidies (even though this is a multimillion-dollar industry). As a consequence, the guide is riddled with contradictions and quite obviously the implementation of its programs cannot be impartial. Take for instance the breakfast and lunch school program—it is for the benefit of the public, but it must also benefit the cattle industry. Who wins? It is essential to cut any ties that exist between the departments of government and private industries.

Another valuable public program is the women infants and children assistance or WIC, which distributes fruits, beans, peanut butter, grains, cereals, fish, and vegetables to six million people monthly. The program must hand out the subsidized milk, eggs, cheese, and juice. This keeps such industries artificially afloat with undeserved assistance.
Did you know that the same system that indoctrinated physicians, also enrolls hundreds of patients in their classes, to teach them how to manage their diabetes? These programs perpetuate incorrect notions, such as: the person with diabetes should not eat too much fruit; once you have diabetes there is no way back; corn tortillas make sugar worse. The focus of these programs is to teach you to manage the disease, not how to revert it. Participants learn how to count carbohydrates, balance medications, and share "healthy" recipes of cheese and chicken, which keep them insulin resistant.

Diabetes is just one example. Multiply this for every specialty that works similarly with its pet disease and the result is today's medical-industrial complex. The truth is that 80% of the chronic diseases are different metabolic manifestations of the same diet and lifestyle.

It stands to reason that our family genes contributed weaknesses and strengths that channel expression of the characteristics which differentiate diseases, but they brew in the chronic inflammation generated by the same diet; we have the finger on the switch.

While it is true that 80% is not 100%, when we choose the diet of our species, a lot of suffering is avoided. There is no different diet to treat each disease. There is one healthy diet for the human being. Eating mostly whole plants low in fat, the majority of us could retain the free state of health with which we came to this world.

Today, it is impossible to imagine a world without propaganda messages on radio, television, the World Wide Web, the expressway, or social media repeating that we deserve free access to medications in order to live. However, before 1997, it was illegal to advertise prescription only drugs; the single other country that allows this today is New Zealand. In truth, most of us could expect to live full lives without the need for prescribed medications.

that would look to guarantee universal education, public transportation, access to disease care, day care centers for children and the elderly, and a clean and safe environment. In order to live free, we need to create the conditions that would permit us to make decisions that could benefit us, one truth at a time. We pay taxes so these and other services can stay within everyone's reach.

The idea of privatizing our public services benefits oligarchs whose purpose is to enrich themselves and their stockholders. Try to think of examples of private companies that offer better service than when they were in government hands (potable water to homes, reliable electricity, protection of coastal waters and reefs, dependable mail delivery to every corner of the world for pennies) and that do it without exploiting the worker. They don't exist.

If you are showing contempt for what the government does and exalt the work of the private sector… think again. Advances in science, medicine, electronics, engineering, transportation, aeronautics, health, education, agriculture, construction, and new materials that are used in the private sector, began with funds from the government before they transitioned into private hands. All of which private enterprises have been able to develop, was initially created with government initiatives and economic push.

Many inventions arise during war time, when the government channels funds toward certain projects that would give it a winning advantage. During peace, private universities lend their professionals, buildings, instruments, equipment (that only their endowment is able to finance) and with government grants (a proportionately smaller investment) small enterprises carry out investigation that benefits humanity.

We must keep an eye on the management of public services. They belong to us. They don't take care of themselves, it's up to us to keep and maintain them. If we allow them to fall in private hands, they cease to be ours. They will exist for the purpose of making money for some, at the expense of the excellence of services that we all need. Their governing boards are in the hands of men (I say men on purpose) that see everything through a money-making lens for them and their circle.

Private corporations erect empires with cheap labor and then put their names on the façade of music halls, universities, investigation centers and golf clubs, as if the accomplishment was theirs.

Today more than ever, medicine has access to a huge amount of drugs for the management of symptoms. In the three years of my residency, the list of medications to manage hypertension grew from three to fifteen. Physicians stopped going into the community. They made people come to centralized places to be diagnosed and have their ailments managed. This convenience gained popularity with providers of the service.

Medicine grew more distant from the communities. It got disconnected from the air, the water, the land. Blind to where people were getting sick and without seeing what kept them healthy, it neglected prevention, and had not much to say about the effects of a protective exposome. Corporative special interests encouraged the unrestrained consumption of processed foods such as cheese, fritters, sodas, and pastries on top of living before screens that stay turned on twenty-four hours a day. The public, too, forgot to create conditions for health.

Isolated from the circumstances that provoke disease, medicine adopted a reductionistic focus in which the organism was replaced by molecules, and the whole could only be handled in fractions. The cost of operating within such a system is exorbitant and the results are very poor, but the house that prints money has grown too big and pays many salaries.

LIFESTYLE MEDICINE

In 2004, lifestyle medicine was born. It is the fastest growing medical specialty, becoming popular among health professionals and patients, alike. Apparently, its time has come. Lifestyle medicine centers on the prevention that is capable of avoiding, managing, and reverting about 80% of chronic illnesses. The American College of Lifestyle Medicine (ACLM) is a 501(c)3 not for profit organization with more than 13,500 active members and a growing number of medical residencies throughout the country, promoting a new generation of physicians educated in nutrition based on whole plants.

In fact, few physicians encourage a diet of whole plants low in fat or guide their patients on a healthy lifestyle. They do not elaborate on the importance of keeping a normal weight, living with purpose, restorative sleep, without toxic stress, and free from violence, conditions that allow us to live many years without disabilities. Perhaps, because few of them live that way.

The medical industry has narrowed its focus into amassing material wealth with screening tests: mammograms, PSA (prostatic specific antigen), bone density, blood sugars, and health fairs where they can identify more people with high blood pressure, elevated cholesterol, or some other condition. Rather than assure that we eat the five portions of vegetables and fruits recommended by the same government and having us avoid processed foods, medicine reduces the importance of risk factors in the community and pushes its laundry list of tests, with hopes of catching more subjects for treatment.

Here I am speaking of medicine in the United States, hopefully some of you readers have different experiences. It will only change with your help. We must educate ourselves and our community, one truth at a time.

MEDICAL EDUCATION

The absence of nutrition education created space for industries to indoctrinate our students. Many physicians ignorant in nutrition repeat propaganda messages, like that cow's milk is the best source of calcium, or that animal meat is our source for protein, or that colon health depends on fiber powder. In doctors' offices, there is no lack of "educational" pamphlets with corporate logos.

Absent prevention tools, medicine presents detection tests as lifesavers, when they really channel people to the micromanagement of diseases. I am ashamed to say it, but medicine has perfected a system in which success of a practice is measured by how many sick patients the doctor sees. The true success is in how many patients improve their condition and transition to become healthy people.

DRUGS AS THIRD CAUSE OF DEATH

In 2023, behind heart attacks and cancer, the third cause of death in the U.S. and in Europe, were pharmaceutical drugs. Double edge swords must be used very carefully.

Anti-inflammatories:
If you are taking drugs to lower pain due to inflammation, you are also reducing the ability of your immune system to react against a virus, bacteria, or a foreign body, when you need it. Did you know that people waiting for an organ transplant have their immune system incapacitated on purpose for a while, to keep them from rejecting the new organ? During that time and in that condition, a person cannot defend even against the common cold.

Nitric oxide:
Nitric oxide is a gas that relaxes arterial walls. We make it in our own bodies using the nitrates we obtain from plants in the diet. Long-acting nitroglycerin for chest angina and the prescribed sildenafil for erectile dysfunction are both nitric oxide donors that relax the arteries. If these two drugs are used simultaneously, their added effect could have grave consequences.

We want the nitroglycerine to dilate or expand the coronary arteries in order for them to carry more oxygen and nutrients to the heart tissues; the purpose of sildenafil is to dilate and expand the penile artery which in turn can collapse the accompanying vein and cause an erection. Once inside the body, these drugs (one for impotence and the other for angina) will relax all blood vessels indiscriminately. Their added effect is capable of lowering the arterial blood pressure to an extreme and cause an ischemic brain infarct or ischemia to the optic nerve that will leave the person with permanent blindness. Your body's production of nitric oxide is wholistic and could never create such double dipping

Insulin:
If you use insulin and by mistake you inject an excess of insulin, glucose rushes inside the cells causing a sudden drop in the blood, which is

left with not enough glucose to feed the cells of your body, and could precipitate a coma incompatible with life.

In this other example, if you are not monitoring your blood sugars, but had reduced your insulin resistance, the usual insulin dose might be too much relative to the sugar in your blood, and have the same disastrous outcome.

When we assume the role of manager of the body, we swim in deep waters. Humbly accept this fact; drugs should be used with extreme caution, supervision, and only when necessary.

> Reference:
> Prescription Drugs are the Third Leading Cause of Death A life without drugs is possible for most of us most of the time. Peter C Gøtzsche, *British Medical Journal* 16 June, 2016.

Every year, the United States graduates 28,000 physicians who are experts in pharma yet unskilled in the role that resistant carbohydrates have on our immune system; on the relationship between cheese and acne, asthma, and cancer; or the fact that we can reverse heart disease.

DISSONANCES

With the discovery of centenarians without disabilities living in the Blue Zones of the planet, many were inclined to wrongly believe that longevity in places like Okinawa was genetic. If that was the case, how can we explain that shortly after the Japanese who move to California they develop overweight, prediabetes, and hypertension, just as Californians? The explanation is simple, they are just like the rest of us, they have no protection against poor nutrition. We find it so difficult to assume responsibility. There is no we and them, it has always been simply we.

Medical schools complain that there is no room in the curriculum to teach nutrition, while according to congressman Michael C. Burgess, MD (R-TX) "there is always time to indoctrinate doctors about the most recent drugs and procedures."

"Unfortunately, physicians have not played a significant role in guiding their patients with good nutrition, because they themselves don't have the necessary education."—Dr. Stephen Devries, preventive cardiologist, who along with Dr Walter Willett, co-directs the Nutrition Education Working Group with headquarters in Harvard Public School.

"Surrounded by light, we stay in the dark." Ryszard Kapuściński, *Travels with Herodoto.*

> Reference:
> McGovern Resolution on Nutrition Education in Medical Schools Passes House; Washington, May 17, 2022

In May 2022, the United States House of Representatives resolved that all institutions of medical education had to include nutrition in their curriculum, but don't hold your breath. Students must demand it and the faculty must offer it before being mandated by laws or resolutions.

THERE IS ONE DIET

If you do an internet academic search on the best diets to treat diabetes, you will find grains, beans, fruits, and vegetables at the top of the list. However, when it comes to talking to patients, many physicians quickly jump to recommend that they consume "enough" lean protein and suggest even to their patients with obesity that they consume the recommended 35% of their 2,000 calories from "healthy" fats. This fear of falling short of protein and fat is due to an ignorant lack of education. We need to be immunized against disinformation.

If you have come this far not merely turning the pages, you know that the population consumes way more protein than we need. If you have learned the basics about insulin resistance, you know that to revert it, you must keep your intake of fats under 30 grams per day. One 3 oz (100g) portion of lean meat has 22g of protein and 10g of fat. The western diet exceeds the required amount of protein and fat by noon.

The formula for the optimal diet to enjoy energy and health consists of 80% carbohydrates, 10% proteins, and 10% fat. We do not need one diet against diabetes, another one against arthritis, and another one to lose weight; we require the same unadulterated nutrition of grains, legumes, mushrooms, fruits, and vegetables.

According to the mistaken recommendation of 35% fat, people would be consuming twice the limit required to reverse insulin resistance. Let's see (2,000 daily calories x 0.35=700 fat calories divided by 9 calories per gram=77g of fat). Good luck!

As a form of review, let's remember how we arrived at the protein recommendations. During the first year of life, the most accelerated growth period for a human, we do so on mother's milk, whose protein content is between 5% and 8% of its calories. It will take us another 10 to 15 years to reach adult size, let's say approximately five feet (one and a half meters). To achieve this construction feat, we need 10% of our daily calories in the form of protein and 10% in the form of fat. You notice that the percentages added to the 5% and 8% of maternal milk are to insure that no one is left behind, so there is no reason to worry. We need the remaining 80% of calories in the form of carbohydrates to make sure we have enough energy for construction.

The fat in maternal milk makes up between 40% and 50% of its calories. After weaning and during his first three years of life, the amount of fat that a child needs is 30% of the daily calories. Subsequently, the fat requirements drop to about 10% for the rest of our lives. Healthy fats are not found in meats or fried potatoes, nor are they found in pastries or bottled oils. Healthy fats are found in whole foods straight from the ground, such as grains, legumes, vegetables, fruits, nuts, and seeds. This should be easy to remember and even easier to implement.

The amount of protein that a 2- to 3-year-old child needs is approximately one gram per kilo of weight each day. As she grows, and for the rest of her life, she'll require 0.8g to 0.9g per kilo per day. If we calculate the requirements of a 10-year-old that weighs 30kg (66lb), these would be the results: 0.8g of protein x 30kg = 24 g of protein per day; 0.9g of protein x 30kg=27g of protein per day.

If the weight when an adult is 70kg (156lb) the requirements are calculated as follows: 70kg x 0.8g=56g of protein. The average person on a western diet is consuming 120g of protein and 77g of fat! What do you think of these exaggerated amounts?
Aside from proteins and fats, we require a great amount of carbohydrates to carry out all this manufacturing. Let's do the math accepting 2,000 calories a day as the average.

2,000 x 0.80 = 1,600, divided by 4 calories per gram = 400 grams of carbohydrates
2,000 x 0.10 = 200, divided by 4 calories per gram = 50 grams of protein
2,000 x 0.10 = 200, divided by 9 calories per gram = 22 grams of total fat

400 grams of carbohydrates

Oats: 1 cup (25g)
Apple: 1 = 32g
Banana: 1 = 28g
Mellon: 1 cup (15g)
Grapes: 1 cup (29g)
Blueberries: 1 cup (21g)
Rice: 1 cup (45g)
Corn: 1 cup (40g)
Onion: 1 = 10g
Pepper: 1 = 11g
Beans: 1 cup (44g)
Garbanzos: 1 cup (27g)
Mushrooms: 1 ounce (21g)
Beets: 1 cup (17g)
Pumpkin: 1 cup (12g)
Carrots: 1 cup (16g)
Kale: 1 cup (7g)

60 grams of protein

Oats: 1 cup (6g)
Apple: 1 = 1g
Banana: 1 = 1g
Mellon: 1 cup (1g)

Grapes: 1 cup (1g)
Blueberries: 1 cup (1g)
Rice:1 cup (5g)
Corn: 1 cup (5g)
Onion: 1 = 2g
Pepper: 1 = 2g
Beans: 1 cup cooked (14g)
Garbanzos: 1 cup (14g)
Mushrooms: 1 ounce (3g)
Beets: 1 cup (3g)
Pumpkin: 1 cup (2g)
Carrots: 1 cup (2g)
Kale: 1 cup (3g)
Flaxseed: 1tbs. (2g)

28 grams of fat (less than 30)

Oats: 1 cup (2.3g)
Apple: 1 = 0.8g
Banana: 1 = 0.6g
Mellon: 1 cup (0.2g)
Grapes: 1 cup (0.6g)
Rice: 1 cup (1.8g)
Corn: 1 cup (2.1g)
Onion: 1 = 0.3g
Pepper: 1 = 0.3g
Beans: 1 cup cooked (0.4g)
Garbanzos: 1 cup (2.1g)
Mushrooms: 1 ounce (0.3g)
Beet: 1 cup (0.3g)
Pumpkin: 1 cup (0.2g)
Carrots: 1 cup (0.3g)
Kale: 1 cup (0.5g)
Flaxseed: 1 tbs. (4g)
Avocado: 1/3 = 10g

These calculations, mere approximations and reductionistic in nature, illustrate how easily we satisfy our requirements. You should also note that they are amounts that fall within the parameters for the reversion of cardiovascular disease and insulin resistance. There is only one diet.

WE INJECT NUTRITION INTO MEDICINE

The Center for Nutrition Studies T Colin Campbell has graduated over 20,000 people from its Plant Based Nutrition Certificate in English. Since 2024, the course is now also offered in Spanish, with Evelisse Capó at the helm. Check it out. At the time of this writing, the certificate has also been launched in Mandarin and traditional Chinese under the direction of Dr. Jing Zhou. In Colombia, Adriana Cortés MD has developed the program TEN (Transforming your Education about Nutrition), with public access and in which over 2,000 people have participated. We also rely upon a number of colleagues and health professionals worldwide that are collaborating in the education of medical students and residents. There is hope.

HEALERS

In every corner of the world and for millennia, there have existed people who practice healing with teas, herbs, poultices, human warmth, and compassion. They are known by many names: shaman, herbalist, bonesetter, sorcerer, witch, enchantress, healer. These characters are respected and also feared. They dare get close to the sick and touch them. They help people through fevers, poisonings, and accidents; assist in births and deaths; repair bones and hearts. They've always been by our side. What sort of powers might they possess?

> Reference:
> *How Witches Shifted from Daily Healers to Heretics and Dangerous Women Under Christian Rule.* Marion Gibson January 22, 2024

Toward the beginning of the twentieth century, an ascending white and well-dressed elite auto designated themselves the only authority capable of ministering to the sick. The group, who at first had no empiric or scientific knowledge, nor the gift of the healers, advanced with religious endorsement and the support of the wealthy.

The church depended on generous oblations, the ruling class lived addicted to presents from the rich, and the well-off elite found it tasteless to be visited by a plebeian healer. Not much has changed. Together they formed an alliance of convenience and accused healers of being dangerous witches. My grandfather Lorenzo, who arrived in Puerto Rico from Mallorca during the Spanish American war in 1898, would tell us "I don't believe in witches, but they do exist..." I wish I had asked him to tell me more.

Healers accompanied the sick and bedridden, observing up close the most inexplicable human experiences. The early medics had no scientific knowledge, but they sought power. Perfumed and in three-piece suits, they became known for heroic practices with leeches and mercury. With courteous elegance, they killed more people than any healer with their cathartics and herbs.

> Reference:
> *Witches, Midwives, and Nurses: A History of Women Healers.*
> Barbara Ehrenreich & Deirdre English

The modern concept of public health took shape in the twentieth century with the development of vaccines against contagious diseases and the discovery of penicillin in 1928, the first antibiotic that could treat virulent bacteria. The Public Health Department redacted laws to guarantee clean water (1908) and air without contaminants (1963). These public health services are not observed consistently in communities of Blacks, indigenous, and people of color, so we cannot rest.

SCIENCE TOPPLES AND DETHRONES MYTHS

In antiquity, the Mayan god Kinich Ahau shone by day and turned into a jaguar by night; Apolo transported the sun from one horizon to the other in his chariot; and the Egyptian god Ra lived and died every 12 hours. Then in 1851, Léon Foucault demonstrated with a pendulum that the Earth spins on its our own axis creating the experience of day and night.

In the years since I graduated from medical school, we have learned a great deal about the effects of nutrition in our bodies. Thanks to technology, we can explain and replicate observations that until recently were a mystery.

We have discovered that by increasing or decreasing saturated fat inside cells we can provoke and revert insulin resistance; we now know that the young cholesterol plaques are the ones which break in an infarct; and we have witnessed how a low-fat plant diet can resolve arterial cholesterol plaques in a relatively short time. What a surprise it was to find that adipose cells in obese persons suffer from chronic inflammation, which contributes to the pain of osteoarthritis. At the same time, every day we know more about the complex immune and anticancer role of the intestinal microbiota.

Most of us came to the medical profession wanting to alleviate suffering caused by diseases, but then jumped head on to treat everything with drugs, which has not translated into better general health. So many scientific discoveries affirm that we can prevent, treat, and often reverse the most feared chronic disease, by simply altering our diet and lifestyle.

What would happen if we returned to the eating practices of our Latino and native cultures based mainly on delicious and nourishing plants. In the blink of an eye, we can modernize and adapt the dishes to today. The Recipe page of The Center for Nutrition Studies T. Colin Campbell has published for free in English, Spanish, and Chinese over seven hundred plant-based recipes, no added oil.

PUBLIC HEALTH MEDICINE

Public health medicine advocates for the people's wellbeing, supports good nutrition, encourages us to exercise, to satisfy hours of restorative sleep, to seek low levels of stress, ensures we have vaccines against contagious diseases, and fights constantly for the community to have access to air and water without contaminants.

Public health medicine is one of the most important reasons why we must support a government of the people, by the people, and for the people,

In 1508, the Copernicus heliocentric theory explained another phenomenon that had intrigued us for a long time: we orbit around the Sun and not the sun around the Earth as the church insisted.
I mean to suggest by these examples that stories, myths, and legends help us live the inexplicable while science advances. Yet, sometimes we cling to myths with such tenacity that the truth must go through hell and high water in order to come to light.

For example, the relation between tobacco and cancer was denied for a very long time. It took 7,000 studies for medicine to arrive at a consensus. Today, in the twenty-first century, we are denying the irrefutable relation between processed and animal food with atherosclerosis, diabetes, cancer, and autoimmune diseases.

Such industries all use the same tactics to convince us of their innocence or harmlessness, while they line their pockets and drag public health to bankruptcy. We must recognize these maneuvers, no matter how they come dressed. Every era has its own lot and here we will try to dismantle or unrig ours. Living free requires eternal vigilance.

WRONG INFORMATION AND DISINFORMATION

Wrong information is easy to clear up, but disinformation is deceitful and is spread deliberately. Let us begin with examples of wrong information.

Carbohydrates make you fat.

TRUTH: Digestible carbohydrates turn to glucose which we use to generate heat and energy—we will talk about the non-digestibles in the section on fiber. A limited portion of the glucose that we obtain from carbohydrates is put away in the liver and muscles as glycogen, to be converted back into glucose during a prolonged fast. The carbohydrates we obtain from whole foods are not turned into fat. In contrast, excess ingestion of refined carbohydrates triggers elevated insulin levels and raise the production of triglycerides.

Fruits contribute to diabetes.

TRUTH: When the cells of the body fill up with saturated fat, they downregulate or deactivate their insulin receptors impeding the use of glucose, which as you see above is our main source of energy. If insulin resistance persists in a susceptible individual, it could progress to diabetes. Fruits are not a source of saturated fat, so they do not contribute to insulin resistance or diabetes.

Prunes (the dried version of plums) are the indicated fruit remedy for constipation.

TRUTH: Lack of dietary fiber is associated with constipation, which is characterized by hard to pass dry stools or not being able to evacuate regularly. Prunes are rich in fiber and sorbitol, promote beneficial species in the microbiota, and retain water in the intestine. All these properties make them a desirable fruit, but many other fruits (and plant foods) share this profile with the prune. Eating a regular and sufficient variety of whole plants is more beneficial than to only use prunes.

Reference:
The following study was financed by the MIKKI Corp, a company that was launched in 2009 and sells MIKI Prune designed to be consumed alone, dissolved in liquid, spread on crackers, or as an ingredient in recipes.

Prune Juice Containing Sorbitol, Pectin, and Polyphenol Ameliorates Subjective Complaints and Hard Feces While Normalizing Stool in Chronic Constipation: A Randomized Placebo-Controlled Trial. *American Journal of Gastroenterology* 2022 Aug 12;117(10):1714–1717

Our medical care is the best in the world and the envy of other countries.

TRUTH: Although the United States spends around $3 trillion in the management of chronic diseases, when compared to the seventeen richest countries around the world it occupies last place in longevity, deaths due to preventable diseases, maternal and infant mortality, and suicides. So, spending more money than anyone does not necessarily translate into better care. A greater emphasis placed in prevention and reversion of chronic diseases would render this budget much more effective.

Once a disease is declared, there is no way back.

TRUTH: The same diet that protects us from developing diabetes, hypertension, or heart disease, helps us reverse them. This reversible quality of many chronic diseases makes us hopeful that we can recover health no matter at what stage of our lives.

Diabetes runs in families, therefore it is hereditary.

TRUTH: Few diseases are strictly hereditary. What we do inherit are genetic weaknesses and strengths. If we don't put our weaknesses to the test, with risky familial or "inherited" cooking and eating habits, we could aspire to live a whole life without developing 80% of the diseases associated with the western food and lifestyle that run in families, including diabetes.

Athletes need protein supplements.

TRUTH: Athletes who exercise vigorously use up a great deal of energy. If one spends a lot of energy, one must consume a lot of food. Eating more whole food, you obtain more of everything, including the protein necessary to repair tissues under stress because of the exercise. Do not think of isolated nutrients, think of nourishing food.

If you eat only plants, you stay hungry.

TRUTH: Many people, when they eliminate processed and animal foods from their diet, make an ill-informed transition to eating only salads. It is estimated that humans eat about three to four pounds of food per day. You could fill yourself up with three to four pounds of spinach, however that amounts to only 300 to 400 calories (100 calories per pound of greens) which will leave you hungry and malnourished. On the other hand, if you eat three pounds of rice, beans, fruits, and greens, that will easily add up to 2,000 calories, and you will be left satisfied and definitely better nourished.

Bones are flexible.

TRUTH: The mineral inorganic matrix of our bones is rigid. The organic living portion of our bones is made of collagen. Since these proteins are resistant to tension and compression, they protect our bones from fractures.

Olive oil is amazing.

TRUTH: Olive oil enjoys a huge fan base that looks to exonerate it of all evil. Learn who sponsors the studies you are reading. The statement "least bad" should not be misinterpreted as "good" plus we should always ask, "good, for what?" The fatty acids in any kind of oil (but especially in olive oil), besides increasing triglycerides and oxidative stress, cause temporary harm to the endothelial cells, impeding the production of nitric oxide and reducing blood flow.

With these examples I have illustrated that to correct erred notions, all we need is to replace them with factual information. It is quite different and more difficult when we are dealing with disinformation, myths, beliefs, refrains, or advertisements that are disseminated with the intention of winning people over as customers and followers.

The following are examples of disinformation—false information that is intentionally spread to maintain power or control over a narrative, usually for economic reasons. These false notions are constantly repeated with the purpose of running over people's common sense. Consider the following examples and who they might benefit from them.

Cow's milk (or dairy) is our primary source of calcium.

TRUTH: Calcium is an element, cannot be created or destroyed, it only changes owners. The same atoms of calcium have circulated on planet Earth since its humble beginnings. Plants absorb calcium through their roots and incorporate it in their tissues. Herbivores pick up calcium for their skeletons by eating plants. Carnivores obtain their calcium, eating the bones of their prey. Ninety eight percent of the calcium in the bodies of animals is in their bones. When an animal dies, the bones disintegrate with the help of microbes and donate the calcium back to the soil. The cow, the same as every female herbivore of the mammal class, produces milk rich in calcium to sustain her infant, eating only plants. Once weaned, no animal drinks milk as part of their diet least of all the milk from another species. In summary, plants are the dependable source of calcium for all life on the planet.

Children must drink milk for strong teeth and bones (see above)

TRUTH: A full 80% of medical schools don't teach nutrition to future doctors. Physicians who don't find out by themselves where we obtain calcium will repeat industry words like a parrot. Don't be bamboozled.

Milk propaganda is quite the feat. The "Got Milk" campaign didn't pay one cent to its stars, many of them were political figures, sports idols, members of the entertainment world, besides physicians and other health professionals. The dairy industry took advantage of their vanity, of wanting to appear with a white mustache on huge public posters.

Protein comes mainly from meat.

TRUTH: The 20 amino acids found in the proteins of plants are the same ones we use to make our own proteins. The original source of protein is plants.

We should avoid gluten, because it causes all sorts of intestinal problems.

TRUTH: Gluten is a protein only found in wheat, barley, and rye. However, the most common way of consuming gluten is as flour in pastries plus packaged and processed foods. If you avoid pastry, you will be avoiding other undesirable ingredients, such as milk, eggs, butter, and sugar. Has it occurred to you that perhaps gluten is not the source of your intestinal inflammation?

Only 1% of the population has celiac disease, a condition that demands living gluten free (which is exceedingly easy if you follow a diet based on whole plants). The market of gluten free products is valued at more than $6 billion.

Even though only one in ten people have a true dietary allergy, around one in three believe they are allergic to some food. More commonly the problem is a deficiency in microbiota that ferment the foods in question.

References:
Prevalence and Severity of Food Allergies Among US Adults. *Journal of the American Medical Association Network Open* 2019;2(1)

Persistent Economic Burden of the Gluten Free Diet *Nutrients* 2019 Feb 14;11(2):399

Why did the gluten-free craze take off? *ZOE* Updated 17th April 2024

Eating plants is too expensive.

TRUTH: If you add grains, beans, fruits, and vegetables to a diet of meat, chicken, eggs, cheese, and processed food you are already eating, you will obviously spend more money. But the situation changes if you replace the meat, chicken, eggs, cheese, and processed food with grains, beans, fruits, and vegetables. If you buy produce as it is harvested from the ground, instead of buying it triple washed, packaged, and ready to eat, you also save money. Remember also that meat, milk, and eggs are subsidized by the government, while the vegetable and fruit grower receives no economic assistance at all. When you buy vegetables, think of the fact that you are paying only a fraction of what it takes to till the land, sow the seed, and harvest the food. Who has the luxury of selling cheaper?

Calcium supplements protect us from osteoporosis.

TRUTH: It is not recommended to take calcium supplements, since they increase the risk of heart attacks, and also kidney stones and constipation. What we want to avoid is fragility fractures and neither milk nor calcium supplements prevent fractures. Bones are living tissue and to keep calcium in bones we need much more than additional calcium. There is more information about this in the section on bones.

Reference:
U.S. Preventive Services Task Force and calcium supplements.

The best way to protect our health is by having regular screenings (blood tests, colonoscopies, mammograms, DEXA scans, etc).

TRUTH: These tests might detect abnormalities and diseases, but they don't prevent them. The best way to protect ourselves from disease is by following a mostly whole plant diet, and a healthy lifestyle free of toxins and excess stress, that also includes exercise, restorative sleep, purpose, support, and community.

Immunizations are not safe and cause more problems than they solve.

TRUTH: Immunizations or vaccines are the safest, most effective and powerful tool we have to protect the population against communicable diseases. Too many people have stopped vaccinating their children, which has brought back deadly diseases we thought were eradicated, such as measles and polio—if this trend continues, there will be more to come. When the vaccination in an area drops below 90%, the virus moves freely and we see outbreaks as with measles. All the recent victims and deaths have been in young unvaccinated individuals. A lack of correct information makes you easy prey for the voices that spread disinformation and fear.

In 1918 and 1919, influenza caused 675,000 deaths in the United States and between 50 and 100 million to die worldwide. The first vaccine against influenza, which was developed in 1945 in Michigan University, was tested extensively in military personnel to determine its safety and effectiveness before administering it to the public. Today, no more than 50,000 people in the United States and 500,000 worldwide die of influenza. You can see the protective effect that this vaccine has had on the population.

When people remain unvaccinated, they put the population at risk, but especially infants below one year of age whose immune system is still immature. It also puts at risk anyone with a compromised immune system due to illness or age. Our family and friends expect us not to put them at risk unnecessarily.

HOW IS A VACCINE MADE? MY SUPERSIMPLIFIED VERSION

Let's say there is a virus going around causing an infection. Investigators at the Center for Disease Control and Prevention (CDC) extract a tiny piece of the virus and purify it extensively in order to introduce it to healthy people with the goal of having their immune system recognize it as something foreign and mount a defense. That tiny purified piece is called an antigen and the defense that the body makes is called an antibody.

The defense system creates a mold or template of this tiny piece and prepares an antibody specific against it. Understand this, that tiny piece cannot make you sick. It's like using someone's nose so that later you could recognize the person by just seeing their nose. The body now has a template that it saves in its memory and the recipe to manufacture an arsenal of antibodies against that tiny piece, in case it ever encounters it again.

Herein lies the incalculable value of vaccines. If any time later that tiny piece enters your body, it will come as part of the whole virus with the characteristic nose. Now, the virus in its whole form can definitely hurt you, but your system immediately recognizes the nose and activates the production of antibodies that will attack the nose, this time attached to the whole virus, and the infection will be neutralized.

This is the part that confuses people. When you get a vaccine, the production of antibodies requires the activation of your defense system and a whole lot of energy. You know the symptoms, because they are the same ones you feel when your body is fighting an infection: body aches, fatigue, even fever. Except that after the vaccine, these symptoms are not due to an infection and will last only a couple of days. You could dampen them with an antipyretic such as acetaminophen.

Now you know the real reason for how and why you feel the way you do after a vaccine and instead of throwing science in the trash, responding with fear, or joining the wave of disinformation, you can support the process.

References:
Vaccine misinformation and social media. *The Lancet* Volume 1, Issue 6e258-e259 October 2019

U.S. Health in International Perspective: Shorter Lives, Poorer Health *Washington (DC): National Academies Press (US)*; 2013.

At times it seems that the economy depends on our believing these myths. "I will not give in to lies. I will not submit to fear"—Jim Acosta

We have the intelligence and the power to correct disinformation. We isolate the parts to study them, always conscious that there is much more at play than what is in front of us.

Over our long history, we have amassed experience with infectious diseases that have devastated segments of the population. The infectious agents interact with us in a fight for survival, they for theirs and we for ours. Who wins? Well, the more capable one. It may seem simplistic, but humor me for a bit. Let's say there is an agent who is not very capable of penetrating our defenses, we keep it at bay. However, if it is sort of capable and comes in great numbers, like the cold virus, we are likely to lose.

The virus that causes the common cold might not be mortal; it doesn't infect everyone who comes in contact with it, nor affect each of us the same way, but it is ubiquitous. If you try to remember how many times it infected you in the past few years, perhaps you identify times when you had neglected one or more of the pillars of health with poor nutrition, lack of adequate rest or sleep, and undue levels of stress. College exams, a big celebration, the stress of a divorce, or prolonged time caring for a sick relative are situations during which our defense system is being asked to multitask.

The intact skin is a formidable barrier. You could come in contact with harmful microbes and nothing happens. In fact, intact skin is a perfectly good barrier against the human immunodeficiency virus or HIV. However, if you have a small scratch, your skin offers a break or gap through which the pathogen can come in. When I was a child, I swam in a public pool. I remember on one occasion that several other children developed impetigo, a skin infection. Why not everyone? The bacteria was virulent, but it needed a small wound—something which was easy to develop when our skin was constantly wet and brushing against the rough surfaces of the pool.

We inhabit this planet with some serious contenders and scientific advances have protected us with strategies that improve our odds in the event of a dangerous encounter.

Having said that, it has not been easy to deal with an arrogant class of humans. In the mid-1800s the Hungarian physician Ignaz Semmelweis said he discovered the cause of puerperal fever and demanded that everyone wash their hands in the delivery room, but his colleagues refused to believe that they were to blame for spreading infections. The germ theory was finally accepted in early 1900 when using sterile water and washing the hands with soap and water became a requirement and saved countless lives overnight. Likewise, the compulsory use of vaccines made some lethal diseases disappear. The unfortunate return of measles in the 21st century is the consequence of a disregard for the civic obligation of keeping our children vaccinated.

When we are surprised by serious pathogens, such as the influenza of 1919 and the COVID of 2019, we realize the importance of maintaining a healthy immune system and a system of public health that watches after all of us. Millions of deaths during the COVID pandemic were attributed directly or indirectly to the virus. Those with comorbidities and preventable diseases like obesity, atherosclerosis, and diabetes, were the weaker ones and the first to go.

Our dietary choices create the conditions for contagion by virulent agents because it overcrowds millions of animals and subjects the vulnerable population that deals with them to those risks. This fact has not come adequately to light. Don't be content with these few lines, investigate more.

THE RECIPE FOR OPTIMAL NUTRITION

The optimal diet for humans:

> Should contribute enough calories for the daily energy needs,
> Should include all the macro and micro nutrients the body requires,
> Should be supplemented with vitamin B_{12} and,
> Should not contribute to chronic inflammation.

From this viewpoint, a diet low in fat based on whole plants satisfies the strictest nutritional requirements. We have evidence of this in the investigative work of *The China Study* and other observational populations studies such as The Blue Zones.

THE BLUE ZONES: MIRRORS OF LONGEVITY

In 2004, The National Geographic, a not-for-profit society, told the world of groups of centenarians living free of disease or disabilities in places they called Blue Zones (because of how they were ink marked on the map): Okinawa, Japan; the islands of Icaria and Sardinia; Nicoya, Costa Rica; and Loma Linda, California.

They summarized the secrets of the enviable quality of life of these centenarians in the "pillars of health." It so happens that the inhabitants of such zones stay physically active, sleep between seven and nine hours each night, do not use tobacco or consume excess alcohol, avoid other drugs, enjoy low levels of stress, and are guided by a purpose in which love, intimacy, and community rule.

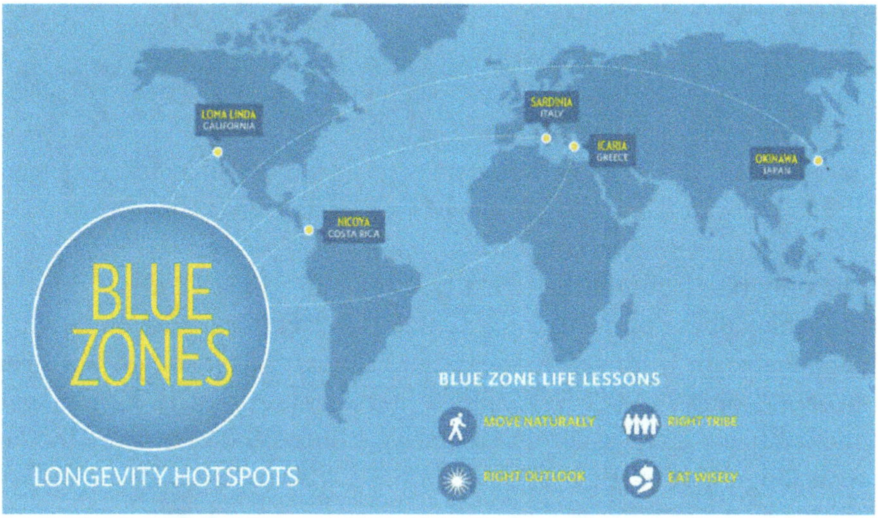

In the Blue Zones of the planet there are many people who reach their one hundredth birthday without disabilities.

These people eat mostly whole plants with little land animal meat or fish. They eat grains, vegetables, mushrooms, fruits, and a great daily amount of beans and legumes. This diet is very nutritious, it is anti-inflammatory, and cultivates a diverse, strong, and healthy intestinal flora. The long lives without disabilities of the inhabitants of these blue zones seem to reaffirm that this type of nutrition is optimal for our species.

The most famous blue zone is Okinawa, Japan. Contrary to speculation, they do not possess an innate or permanent genetic protection, since the Japanese who move to Hawaii, U.S.A. or to San Francisco, CA, soon develop coronary disease just like the longtime residents of the place.

The common factor of these people are not familial genetic ties, but a way of life. This indicates that the most important thing is not the genome, but the exposome or environment to which we expose it. We should ask ourselves why there are not more blue zones.

These are not the only places where people live healthy to advanced age. In places around the world where the fruits of the earth are within everyone's reach, where people participate of nature at all hours, where the wellbeing of the individual is primordial, and where life is in harmony with the planet, people keep their health for many years. We can create more blue zones.

Reference:
Acculturation and coronary heart disease in Japanese-Americans. *American Journal of Epidemiology* 104: 225-247. 1976 Marmot.

The only blue zone in the United States is in Loma Linda, California. One third of the population of 22,000 is Adventist: a religion in which people abstain from eating meat, don't smoke and many do not drink alcohol. Loma Linda is surrounded by communities with high rates of obesity, heart disease, and Alzheimer's. Loma Linda residents live ten years longer than their neighbors.

As in other parts of the planet, the western lifestyle has invaded blue zones in the form of fast food, sedentary routines, less sleep, tobacco, alcohol abuse, excess caffeine, constant stress, a disproportionate emphasis on productivity and an unconscionable exploitation of the individual, leaving no room for people to procure their wellbeing. The singular advantages that identified the blue zones are slipping away.

SUBSIDIES TO THE CATTLE INDUSTRY AND ITS SIGNIFICANCE

Reference:
In 2020, during the peak of the COVID pandemic, the demand for milk plummeted (COVID-19 and the U.S. Dairy Supply Chain published by Agricultural and Applied Economics Association).

For a mega industry that keeps millions of cows constantly producing milk, it was too late to recalibrate. Cows cannot stop producing milk like you could stop the production of automobiles or the construction of houses. Therefore, the industry dumped around three million gallons of milk in the ocean every day. It did not lower prices or give milk away, it rather dumped it.

The production of milk is not regulated by supply and demand. The incentive for the dairy is the government subsidy. So, their objective is to produce and produce, otherwise they won't receive. The Agriculture Law of 1949 established a program of permanent assistance for the dairy industry which promotes dumping excess milk to, among other things, maintain prices stable. The cattle industry exploits millions of sentient beings, sells the meat of its offspring, and throws away its treasured nourishment.

If you drink milk, did you know it came from pregnant cows? Normally a mother produces milk after she delivers, not while pregnant. But the dairy industry is a calf that is never weaned, so as soon as the cow delivers, she is impregnated again to keep the milk production uninterrupted. Several studies suggest a correlation between high milk consumption and poor-quality sperm and a higher incidence of prostate and testicular cancer.

> Reference:
> The milk we drink, food for thought. *Inklings* Volume 106, Issue 6p1310-1311 November 2016

In 1977, an investigative group headed by Senator George McGovern, made public for the first time the problem of malnutrition in the country and launched the original recommendations titled Dietetic Goals for the United States. As a result of this report, every five years since 1980 the United States Department of Agriculture (USDA) with the Department of Health, updates the nutrition recommendations and interprets them in a graph. The depictions which were initially in a pyramid went through several iterations until finally acquiring the shape of a plate, as we will see ahead. Every government and private entity—schools, the military, prisons, hospitals, daycare centers, elder homes—they all follow this guide.

Even though in 1980 there was already enough scientific evidence to recommend we eat less meat and in spite of the preponderance of unbiased scientific evidence associating the consumption of plants with more health and products derived from animals with increased risk of disease, the authors of the guide did not have the courage to make recommendations that would offend the beef growers, so the industry successfully blocked the phrase "less meat" from the guide.

In 2009, the Physicians Committee for Responsible Medicine (PCRM) presented their graph as an alternative to the pyramid. The four food groups (vegetables, grains, fruits, and legumes) are shown in full color photographs on a plate with a glass of water.

The PCRM plate shows the 4 food groups.

It was thirty years after the first guide, when the authors finally had the guts to challenge the industry and publicly recommend that we lower our consumption of animal meat. In 2010, not only did they recommend that we eat more vegetables, fruits, grains, and drink water, but the document also spells out that "we should eat less meat." Decidedly a success.

Don't cry victory, at least not yet. In spite of the step forward, when you actually read the document, its pages are riddled with confusing and contradictory messages, reflecting the powerful influence that the producers of milk, cows, chickens, eggs, sugar, salt, and oils have on our government.

ONE STEP FORWARD AND TWO STEPS BACK

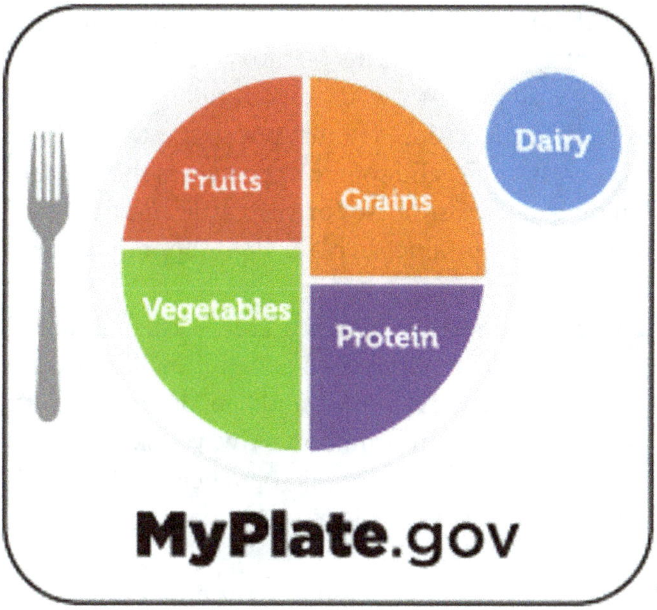

MyPlate, the graphic based on the United States nutrition guide, retains protein as the avatar for meat and advertises cow's milk as if it was a food.

THE CURRENT USDA PLATE

The traditional pyramid is now a plate, which is more in accordance with how people think of food. Cow's milk appears next to the plate as a commercial product advertisement.

The plate is divided into quadrants supposedly representing the four food categories: grains, fruits, vegetables, and protein(?). Protein is not a food, it is a nutrient. As we see, the beef industry agreed to eliminate the word meat and keep its avatar.

You already know that protein is a nutrient made by plants, it is part of the structure of every grain, vegetable, bean, fruit, mushroom, and seed. Animals acquire their protein from plants.

In other words, the fourth quadrant belongs to legumes: beans, lentils, pods, soy, and chickpeas. This group of foods, besides being rich in protein, is chock full of nutrients absent in meat, such as fiber, carbohydrates, vitamin C, antioxidants, and so many other phytochemicals that are found only in plants. Phyto means plant or vegetable.

While the improved graph now appears in the shape of a plate, it needs to clean up its disinformation. It must not suggest that cow's milk is an essential food or that the valid source of protein is the ghost of meat. The nutrition guide is a Trojan horse for the cattle industry where it hides the propaganda for its product in plain sight, hoping we don't notice.

The economic relationship between the cattle industry and the department of agriculture reflects a shameful conflict of interests. Join together in dismantling the propaganda in order to substitute it for independent and truthful information. Always ask yourself if a person, entity, or organization is its economic beneficiary. Pull the curtain and expose the Wizard of Oz with his megaphone.

"We shall not cease from exploration, and the end of all our exploring will be to arrive where we started and know the place for the first time."

—T.S. Eliot

The principal objective of a for profit corporation is to turn everything into money for its stockholder elite. The role of the "consumer," that's how they call us, is to consume. Even the most educated person can have the mind colonized with these ideas and participate in their own destruction.

I am hopeful that as you discover the marvel of your body and the plants that nourish it, you will make decisions based on truths and not propaganda.

We need the truth just as the truth needs us. With this infoxication mess, the most obvious victim has been our body. We underestimate this immense treasure that communicates with us through thirst, hunger, sleep, satisfaction, pain, fever, sadness, and love. We flood it with drugs, subject it to invasive procedures, and torture it with diets.

Whenever we feel sick, we look for something to take. Rarely does it occur to us that we might need to eliminate what might be hurting us. Try to come out of a drunken state by drinking coffee or splashing water on your face, we know very well that you will be under the influence of the alcohol until the body metabolizes it and the levels drop in your brain. The same happens with inflammation, until the body is no longer under the effects of inflammatory cytokines and prostaglandins you will feel pain and discomfort.

SUPPLEMENTS FOR PREVENTION

The practice of taking supplements indiscriminately, in case the body needs them, is not only a waste of money but clutters us with substances that the liver and kidneys need to eliminate plus increases the chances of grave interactions and side effects. If this was not enough, when these substances are excreted, they contaminate the planet's great bodies of water, since not all of them can be eliminated by the water treatment plants.

Besides, these ingredients that we call supplements were part of "wholes" in nature and interact with others to complete biochemical reactions in the bodies of animals. We never ingested them in isolation. When they come by themselves or in bad company, they do not behave as they should and occasionally, they have the opposite of the desired effect.

Science has enabled us to isolate hundreds if not thousands of ingredients from vegetable and animal tissue. This means we have the technology and the instruments to separate the whole into its fractions to study the characteristics of each ingredient. For example, we can separate the bran, germ, and endosperm of grains. The people who work in such projects are given awards and recognitions, because they help us understand the world in which we live. Having said this, we should be aware that each separate part does not function as in the whole!

Nonetheless, once isolated, the temptation to monetize it as a "new product" is very strong. Someone will find a market for it, even without merit, and propaganda will do the rest. That is how we fill pharmacy and bodega shelves with ingredients that we could better obtain by eating the whole plant.

Some examples are psyllium (a fiber that is in all plants), probiotics, (bacteria that we swallow with the food and take residence in our gut, calcium (an essential mineral present in all vegetable tissue), and collagen (proteins that each animal makes according to his needs).

SUPPLEMENTS ARE ISOLATED FRACTIONS

In addition, and as I have mentioned, these fractions do not behave like the whole. The following study is widely known for its dramatic results. It had long been observed that smokers who consume foods rich in carotenes enjoy a certain protection against lung cancer. Extrapolating this notion, investigators in Finland designed a study whereby carotene was supplied as a supplement to people who smoke. The results were so unexpected. It not only did not protect them, but instead they saw an increase in cancer and had to stop the study. When we use isolated ingredients, even when they have been extracted from plants, the results are unpredictable.

MODEL FOR OPTIMAL NUTRITION

Our ancestors were sustained by food they found along their way, food that did not run, fly, or swim away. For hundreds of thousands of years, we ate what we could reach with our own hands. We descend from humans who reached reproductive age by eating mostly plants.

The environment in which we developed lacked too many edible plants. So, it is logical to think that we would profit from unattended eggs, hurt animals and some unfinished prey of carnivores. There is archeological evidence that we were butchers before we were hunters, which allowed us to increase our calorie intake, live a few more years, and leave descendants.

Reference:
How Hunting Made Us Human —*Scientific American* April 1, 2014 About 279,000 years ago in a mountain range near a lake in Central Ethiopia, it seems that the first hunters created sharp objects out of black-green volcanic glass.

To prosper in our modern era, we must consume enough calories that we could convert into energy, together with the macro and micro nutrients that we need to grow, maintain, and repair our tissues and humors; we must also use a vitamin B_{12} supplement (absent in modern diet) plus keep the body free of chronic inflammation. The type of diet that meets all these requirements and confers humans more life years free of disabilities consists of an abundant variety of colorful whole plants every day.

In spite of knowing all of this, how many humans keep gallons of pasteurized milk in the refrigerator, cheese made to the tune of one pound per seven of milk, buy trays of plucked chicken breasts and breaded wings, farmed fish, frozen shrimp, ground beef, quartered turkey, sliced ham, eggs by the dozen, novelty pastries, bottled oils, sweet drinks, caffeinated water, and manage it all without stepping on actual soil.

The way to maintain a population that eats mostly animals and processed foods is with guaranteed government subsidies, allowing the use of tons of chemicals (many of them toxic), securing huge land masses, employing armies of machines, and exposing workers to risky conditions. The lagoons of excrement from the concentrated animal feeding operations (CAFOs) where animals are kept confined, often overflow contaminating the surroundings and placing us all at risk. A diet of meat is costly and unsustainable.

All this despite the fact that the planet already harvests enough corn, wheat, oat, rice, bean, lentil, soy, garbanzo, sweet potato, potato, pumpkin, cabbage, cauliflower, spinach, mushroom, apple, jackfruit, yuca, blueberry, walnut, cashew, almond, flaxseed, sesame, and chia.

If we chose food directly from the ground instead of passing it first through the stomach of an animal; if we ate corn, soy, wheat rather than wait for the animal to turn it into muscle for the butcher shop, the calories and nutrients would serve us a hundred times better, with the advantage that there would be enough food for the world population, today! If we practice what we are reviewing here, the world could serve us all.

Flexibility in eating. The human digestive system specializes in decomposing carbohydrates to turn them into glucose, which is our principal source of energy. In addition to digesting plants, we can metabolize animal tissue and factory food. The domestic dog evolved alongside us eating and also digesting most of our food, including plants; my Perla likes small pieces of raw cauliflower, baked sweet potato, banana, and mango. Much like we do, dogs have certain omnivorous flexibility.

On the other hand, obligatory carnivores lack the enzymes to complete the digestion of carbohydrates in plants and have less flexibility. They feed on other animals that used vegetable protein and made the muscles of their bodies. If you have stumbled upon the chalky excrement of a coyote or a fox, it is due to its calcium content, which indicates they recently ate bones. The mountain lion satisfies his energy requirements with 3,000 calories a day of protein and fat, the equivalent of 10 pounds of meat (about half a turkey). However, his brain depends on glucose, so he uses gluconeogenesis to convert some proteins into the glucose required.

Autotrophic. Thanks to photosynthesis, plants take carbon dioxide and water and convert them into carbohydrates, proteins, and fats. They make themselves and the food upon which all life on the planet depends. What a marvel! Yet for us animals, eating is a requirement.

Plants are the most accessible food (easy prey), since they don't run, fly, or swim. Plus, how amazing that plants are the only source of carbohydrates, the nutrient that we readily convert into glucose to generate most of our energy. (Let's not forget the resistant carbohydrates discussed in several other sections.)

According to very valuable archeological findings, when humans had the opportunity, they supplemented their diet with animal remains. Imagine our ancestors walking in small groups from here to there collecting everything they could eat and store, trying what they found along the way—a juicy root, a poisonous one, a sour and fibrous fruit, another one ripe and sweet, a new seed and a rancid one, an unattended egg, the leftovers of a recently abandoned prey. They were the first scientists, observing and testing whether to eat it or store it and for how long—I imagine those as our beginnings. Why are we told the myth that because of their curiosity a famous couple was thrown out of Paradise? Curiosity has been our faithful companion, always.

It was advantageous to consume some animal meat. It makes perfect sense, since besides protein it contributes a generous portion of its calories in the form of fat that could be stored for later.

In these modern times besides plants and animals, we have a third source of food: factories. In their laboratories we find fractions derived from plants (flour, oil, and sugar) combined with salt, milk, cream, cheese, egg, meat, condiments, preservatives, flavorings, and more—"novelties" that never existed in nature, such as breaded ham, sausages, chocolate sauces, attractive glazes, sweet donuts, liquid sugar, plus a large combination of addictive spices added to the elaboration of double cheese pizza with pepperoni, fried corn pockets filled with ground beef, nuggets of chicken and turkey, an immense variety of sweet and salty fritters, and so on.

It is essential that we understand this group of edibles of high caloric density, if we want to understand the obesity epidemic.

TRUTH OR FICTION IN NUTRITION

The term "CARBS" originated in the 1980s as an abbreviation of carbohydrates. Today it has become synonymous with flour and sugar, the carbohydrate fraction that has been isolated from the plant. However, we do not eat flour or sugar with a spoon directly from a bag, so when people say CARBS, they are really referring to bread, pasta, crackers, cookies, donuts, pastries, and cakes elaborated with flour and sugar.

In contrast, a diet with a variety of whole plants rich in carbohydrates is associated with health, while the western diet rich in "CARBS" carries a higher risk of death from all causes. We make a fundamental mistake when we confuse the carbohydrate isolated from the plant with actual plants rich in carbohydrates, such as rice, oats, sweet potatoes, potatoes, pineapple, and corn tortillas. Have you stopped eating whole plants rich in carbohydrates thinking they are "CARBS"?

Cow's milk enjoys undeserved fame as the "only or best source of calcium." This is a great lie perpetuated by the dairy industry. Plants absorb calcium from the soil through their roots and incorporate it in their tissues. Animals obtain the calcium to fabricate and maintain their skeletons by eating plants. Plants are the source of calcium on which the female mammals depend to make calcium rich milk for their young. Do you fill your refrigerator with milk, cheese, yogurt, and ice cream based on a false premise? Think of the consequences this disinformation has for you and your family. If you lead an organization, multiply the effect of applying this disinformation by every member of the group. And if you are a government official, what might be your effect?

On social media (and elsewhere) there are people saying that, if you do not eat animal meat, you will be lacking in protein. In fact, the 20 amino acids that make up plant proteins are the same 20 that every other animal on the planet uses to make their proteins, including us. So, we'd be lacking nothing. During the first year of life (the most accelerated growing period of a human) we grow with maternal milk whose protein content is between 5% and 8% of its calories and we reach adult size with that same amount of protein from our food, between 5% and 10% of the calories per day. To leave you no doubt, 100 calories of broccoli, lentils, or soy concentrate the same amount of protein as 100 calories of beef: 10 grams. Protect yourself with the truth.

You might have fallen prey to the many stories about olive oil and to the obsession with healthy fats. If you look around, it is unlikely that you will spot even one person deficient in fat. Oil is the most caloric fraction of the plant: 4,000 calories in one pound. Some physicians quote studies designed by the olive growers and recommend that you drink it daily, without taking into account the negative effect that oil has on weight, on insulin resistance, in the endothelium and on the circulation. These recommendations don't stand on a firm scientific basis. Don't fret, cooking without added oil is not cooking without fat. What it means is that you will take advantage of the fat in the whole food—garlic, onion, nut, seed, avocado—instead of adding it from a bottle.

Don't fall for the promotion that you are saving time by stocking your kitchen with energy bars, granolas, bags of nuts, boxed cereals, precooked meals, yogurt cups, cheese bites, Lunchables and other easy to carry packaged snacks. Industries release 4,000 calories daily per person, that is 2,000 more calories than required by an average adult. If you have gained weight or can't lose it, don't blame your genes. Feed your curiosity with truths.

PART FIVE

How I Landed Here

Last night someone asked if I had always been serious and studious, if I have children, why since I'm a doctor I cook with the community, why I don't charge.

MY TRAJECTORY—SORT OF

It has helped me to believe that we come from something BIG. That when we are born we leave it behind. It animates me to think that when we first open our eyes to this world we begin to remember day by day what we once knew without understanding it : a likeness of that memory imparts certain familiarity to the journey. Emily Dickinson said, to live is to wake up new every day. I imagine that in the end, we will return to the immensity, more conscious of the mystery.

I sampled being timid and also funny, until I found my voice. Before I deserved it, my mother proved to be my unconditional fan. It was easy to grow under her wing. One rainy afternoon, I was forbidden to step outside. I put one foot outside. Tested my autonomy. I explore new things. I learn from experts and novices, and return the favor. I am loyal to the truth. I embarked on decades of psychoanalysis and discovered truth in the lyrics that say that when I turn out the lights, I may not see what surrounds me, but I know it's mine. Curiosity has not killed this cat.

BEGINNINGS

I was born and raised in America, Puerto Rican, Boricua, mother, daughter, bilingual, teacher, writer.

As a little girl, I would spend hours stretched out on the cool tiles: face to face with prehistoric lizards with a dorsal fin, three lids, and ears without auricles. Upon yawns they let me see their bright red tongues and miniature toothless mouths. When they pushed up on their arms, they brandished a menacing blade in the neck. High up on the walls, their cousins the salamanders wriggled in the moonlight. Their translucid bodies provoked admiration and apprehension in me.

Our front porch cool tiles were my box seats to the world of which we are a part. Spread over them I was one with nature.

The street vendor walked from the farmer's market in Río Piedras. Don Guille parked his cart loaded with plantains, pumpkin, sweet little peppers, beans, and more right in front of our house. He would give me a fistful of white rice that I pretended to cook with water, salt, and broad leaf cilantro in my small aluminum pot while my mom prepared lunch. I was not four feet tall when we got a Bendix washing machine with two rollers that could squeeze the water out of clothes. From the back patio one could access three doors; the first led to a bathroom, the second to a service room, behind the third and last was my dad's photography dark room. Standing on a soapbox by the light of a red bulb, I saw images revealed in vats of stinky chemicals. My dad taught me to always look out for nature. We fed flocks of orphan birds with an eye dropper and saw each of them take flight. In the kitchen with doña Juana, I learned how to quarter a chicken, the irony.

I used to leave fruits and vegetables in the spaces of the cement block fence as offerings for sidewalk passersby. At night, my mom set up her Singer and clothed my two sisters, me, and also her.

I talked with the coquís (tree frogs) that sang beneath my window. Not infrequently a mosquito snuck under a corner of the net over my bed, her humming woke me up. After sunrise, I traced the shiny trail left by snails and slugs on the tile walk outside. The song of birds, the crow of roosters, and the noisy garbage truck signaled daybreak. I absorbed the languages of my surroundings.

When I began school, so much of what I wanted to explore stayed outside. During my teen years I feigned being docile and tame. Thanks to the Wild Woman, I never totally submitted. In high school when my peers asked nonstop, I paid attention. Once in University, I met courageous students who had moved far away from their families to study in the city. One good day, I would imitate them and take flight.

> Reference:
> *Women Who Run With The Wolves* by Clarissa Pinkola Estés, PhD

EYES OPEN TO THE OCEAN

The medical school stood right before the Atlantic ocean. By the sound of waves breaking on the rocks, we unnecessarily broke dogs and rabbits. We learned anatomy from an ageless cadaver. We yearned to know patients.

At last, we were transferred to the Medical Center, a hospital with open windows, where patients drank their cocktail of drugs and bathed in sun balconies overlooking the gardens. I will never forget such a postcard scene.

Six-year-old Daniel suffered from glomerulonephritis. Don Rogelio, scared and alone lay in the intensive care unit. Alma walked the corridors day and night. I lost my innocence listening to their stories.

The patients' lunch and supper with a glass of white milk on the tray smelled always the same. We were graduated without ever hearing the role that food played in health or disease.

FIFTEEN HUNDRED MILES

I never knew the international world as closely as I did during my internship and residency: Rojas, Shah, Ramachandran, Nguyen, Lee, Patel, Johnson, Negrón. I had dedicated professors and studious colleagues. We saved some lives with antibiotics, while the list of drugs against pain, inflammation, hypertension, insomnia, and depression grew with each passing year and with them the secondary or side effects we should learn to suppress with more drugs.

Down the halls of the hospital, we crossed paths with well-dressed young counterparts that in their wake left pens, magnets, mugs, information sheets, every now and then a book, and scores of trinkets each with an emblazoned logo. They came bearing gifts for us in the form of pizza, donuts, sodas, and coffee. Pharmacy reps, were they. In time I came to understand this pact of convenience and more.

They talked to the attending physician of this or that new drug. Showed up with samples requiring their signature, bottles with enough pills to start a patient's treatment of hypertension, cholesterol, depression, or some infection. Between having something free to give the patient and the time required to learn from independent sources plus a measure of intellectual laziness, all of this paved the way. We would remember the hand that fed us.

This relationship grew unchecked. When the office suddenly wished to require them to make an appointment before meeting with the physician, it was too late. They not only brought lunch to our conferences, but also organized continuing education opportunities over steak and lobster in nice restaurants, gifted tickets to events and concerts, knew the names of the doctors' children, the wife's preferences, the husbands' sports affiliations. (It was the sexist 70s).

I avoided them at all costs. Not out of arrogance, but as a precaution and protection before my admitted weakness. I felt that if I exposed myself to their sweet talk or accepted their gifts, I would be incapable of prescribing with impartiality.

Pharmacy contributes enormously to our emergency medical care, but we have been indoctrinated to apply these remedies to a plague of chronic diseases that stem from our food and lifestyle choices. Such application of medicine is inappropriate, costly, and ineffective.

The following words from president Abraham Lincoln stay with me: "At what point then is the approach of danger to be expected? I answer, if it ever reach us, it must spring up amongst us. It cannot come from abroad. If destruction be our lot, we must ourselves be its author and finisher. As a nation of freemen, we must live through all time, or die by suicide."

1957 SET MY PATH

The rainy season in Puerto Rico was accompanied by swarms of mosquitoes where in any dark corner females turned to blood sucking vampires. I remember the first of many days the "tifa" or DDT (dichlorodiphenyltrichloroethane) pesticide spraying truck came to our neighborhood. We had to close our windows and flee from the house to my grandmother's. I don't know what my sisters thought, but for me it was terrifying. I didn't know if we would be coming back. Eventually the cloud settled. For days all smelled of poison. Some years later, in 1962, Rachel Carson exposed the atrocious consequences of this lethal chemical in *Silent Spring*. The eagles produced eggs with shells so thin they broke when the bird sat on them—the world lost more than two million raptors. This was my entry ticket to activism.

IN SCHOOL

When my son was in elementary school, we led an awareness campaign on recycling. It culminated in a student made sculpture with home-found plastics. It was left in the foyer for several years sparking more environmental campaigns, some remember it almost forty years later.

AT HOME

The night I finished reading *Diet for a New America* by John Robbins, I announced to my family that we would bring no more meat into the house. I traced a line in the sand and to my surprise there were no objections. Jason asked me, why exactly are we doing this? We were in the habit of sitting at the table as a family for breakfast and dinner, moments during which we shared dreams from the night before, events of the week, anecdotes, interesting learnings and discoveries. That night I didn't need to say much, arriving at the top of the stairs, he exclaimed that's good enough for me and to my knowledge has never looked back. The following week while playing at a friend's house his mom called to ask what she could give this self-proclaimed vegetarian. Lack of imagination is often our biggest obstacle.

Years later as I invited colleagues to discuss the science of nutrition it intrigued me that the more educated ones were the least interested. Those who have dedicated years of study reinforcing ideas to which they've subscribed—that calcium comes from milk, that animal protein is the one, that diabetes is incurable, that heart disease can only be managed—are afraid to burst the bubble in which they live, they fear the house of cards would collapse. Many are still paying the loans for their education.

On the other hand, people most open to learning have nothing to lose but their ignorance. The most interested in learning have not erected a lifestyle fed by ideologies and myths or beliefs, so they can let go of false notions with more ease.
A very reluctant group includes not just health professionals such as doctors, dietitians, and trainers, but also people who have defended these ideas before relatives and friends. It is very difficult to open ourselves to new knowledge if we must accept having lived according to a false

concept. In order to save face, not be humiliated or lose prestige, we are capable of defending a lie.

> Reference:
> Go to the section where I talk about the five aspects of our personality.

When we can stop on a dime and consider new evidence independent of economic interests, the path does not close. Just the opposite, it opens up with innovating questions that were not asked before. How is the milk of every mammal enriched with calcium? Does it matter which source of amino acids we use to make our proteins? How do the nutrients we consume as supplements behave compared to the nutrients we consume while still in the whole food? If it is not inevitable to get sick as we age, do you feel enthusiasm or bewilderment? It appears that this natural complexity terrifies too many people who would rather live by simple and immutable answers.

ONE WORLD CLOSED AND ANOTHER OPENED

When I recognized the demands of my palate, it revealed a violence that caused me remorse. The destruction and waste were repugnant. From that day on, I stopped eating the meat of chickens and cows, drank milk no more, afterwards stopped eating cheese, gave up eggs, and eventually fishes. With my heart in pieces, I continue to try and make peace with a universe that offers me more than what I ever imagined.

IN MY PRACTICE

Already board certified in family medicine and just like my colleagues, I prescribed drugs for diseases. I was ignorant about nutrition and even more so when it came to the topic of nutrition without meat, eggs, or cheese, since my decision was not based on science, but rather philosophical and ethical. Besides, my patients are not my family and my consultation room is not my kitchen.

I felt an immense curiosity. Little by little, as I found scientific evidence and respected medical professionals expert in this lifestyle, I forged the bridge toward a state of health based more and more in whole plants. My colleagues took every opportunity to point out that by questioning the medical paradigm that had formed us, I was making a mistake. Some were more frank and reduced my rank by saying that I did not practice the same quality of medicine they did.

A SABBATICAL

After a decade of strict medical practice, I decided to dedicate some time to study the relationship between nutrition and health, but free of economic conflicts. Curiosity carries risk, so I climbed down from the ivory tower (that's what we call the silos of academia) and paid myself a two-year sabbatical.

I collaborated with *The Vegetarian Resource Group* in translations. Became a member of Physicians Committee for Responsible Medicine. I attended conferences from coast to coast. Read studies and investigations coming out of Loma Linda, the renowned Adventist Medical Center. I learned of the lifestyle associated with longer years of life without disabilities. I gathered enough concepts and confidence to start practicing a medicine which confers these benefits, a medicine based on whole plant food nutrition.

I treaded lightly, thinking that I was distancing myself from conventions and that I might end alone and marginalized. Fear and ignorance have always been used to control us, but love of life guides our progress, so I kept going.

I enrolled in more nutrition courses based on plants and learned from experts; I met curious physicians and investigators that would not stop asking questions like why the modern diseases are on the rise; physicians interested on why the incidence of osteoporosis varies around the world; health care providers that instead of wanting to change the behavior of children with obesity (some protocols suggested putting a padlock on the refrigerator door) saw an environment that laid a trap which few escaped on their own; doctors that lifted their gaze to recognize that in places where addictive food was not yet ubiquitous, people retained their health and enjoyed more years of life without disability; physicians that were not loyal to industry.

I studied under Antonia Demas, PhD, whose doctoral dissertation *Food Is Elementary* won the most innovative use of the food pyramid (the food guide for Americans) awarded by the USDA. Did they realize the prize was being awarded to a curriculum based only on plants without animal products: meats, cheese, or eggs?

Within a short amount of time, I met colleagues who wore no restrictive blinders or pharmacy-colored goggles. In two years, I had filled my medicine bag with solid concepts of this effective and powerful medicine. I dared to prescribe a whole foods plant-based diet with a new purpose: try to prevent, counteract, and place in remission some of the most feared and costly diseases of our time.

Rather than convince participants of my presentations to adopt this diet, I wanted them to experience the connection between food and wellbeing for themselves. I began asking with purpose and insistence, what is in your refrigerator, freezer, pantry, and cupboard? That is how the Virtual Kitchen Tour was born, I could not have learned this information any other way.

It surprised me then and still now, how so many adults justify the presence of junk or unhealthy food in their homes saying it is for their children or grandchildren. They know it is not quality food, since they themselves often report not drinking cow's milk, juice, sodas or eating deli ham, cheese, cookies, chips, gummies, ice cream, and candy… they are treats for the kids. We have evidence that arteries begin to clog at an early age with this diet. How can people maintain a dissociation between what is good for them and what starts disease among the young in their family? Aren't they ashamed? Or is it that their brains are so awash with disinformation that they cannot register dissonance or feel curiosity?
According to autopsies of soldiers whose average age was 20 years and who died in battle during the Korean War of 1950 to 1953, 77% already had atherosclerosis in their arteries.

Reference:
Manifestations of coronary atherosclerosis in young trauma victims—An autopsy study. *Journal of the American College of Cardiology* Volume 22, Issue 2, August 1993, Pages 459-467

In 2005 I read *The China Study* by T Colin Campbell PhD and for the first time stood on firm ground. Since then, I have read it several more times (now on its second edition). This study offers an immensely valuable frame of population observations. The international experts exposed some of the most pertinent nutrition investigations of our times. I beg you, do not criticize it before reading it. Do not accept reviews of others who haven't read it. Be skeptical of the opinions of those that want to advance ideologies instead of truths.

Reference:
On Tyranny: Twenty Lessons from the Twentieth Century by Timothy Snyder. To abandon facts is to abandon liberty. If nothing is true, no one can criticize power because there is no basis for it. If nothing is true, then everything is a spectacle. He who has the most money, pays for the brightest lights.

LIVING ON PHARMACEUTICALS IS UNSUSTAINABLE

The year Sunstone Press published my first book *Nourishing the Body and Recovering Health*, the United States used $312,000 million on pharmaceutical drugs. Today that amount surpasses $574,000.

We use more than half the gross national product (GNP) in medical services, yet we are the country with the highest rate of obesity on the planet and where the most people live with multiple chronic diseases. Instead of practicing prevention, medicine has dedicated itself to manage the costly consequences of a dietary pattern and lifestyle that have been eroding our health for decades.

It is true that to practice prevention of disease requires effort, but the cost is minimal. On the other hand, treating symptoms and diseases is expensive (or lucrative, depending on your point of view). The annual cost of managing chronic diseases, not counting hospitalizations, is around $3 trillion.

The number one cause of death continues to be heart attacks, a tragedy that can be 95% avoided by eating based on whole plants, following a healthy lifestyle that avoids tobacco which includes exercise, reparative sleep, low levels of stress, clear purpose, love, and community. Hospitals pocket around $123,000 for each bypass surgery performed.

The second cause of death is cancer, which costs the country $190 billion annually. Out of that amount, $4 billion is spent on drugs.

Diabetes contributes to all other diseases (nothing exists in isolation). Some estimate that by the year 2035, the diabetes epidemic will have bankrupted England.

The people who depend on medications to control overweight, diabetes, high blood pressure, and cholesterol represent a potential of earnings for the pharmaceutical industry of thousands of millions of dollars. Therefore, educating people and enforcing regulations that protect public health is akin to killing the goose that lays the golden eggs.

At the same time that we make decisions based on truthful investigations, we must repair the institutions that fail us.

In his book *Whole*, T Colin Campbell tells us of organizations dedicated to publishing true scientific facts.

THE COMMUNITY AND I

Once I found that the nutritional pattern based on plants had a solid scientific base, I began sharing it with colleagues and patients. I told them of the nutrients in food, of the sulforaphanes in broccoli with its protection against cancer, of the amount of calcium in cabbage, and of the negative inflammatory properties of the arachidonic acid found in meat and milk. I saw that patients enjoyed learning, like I did, and was surprised not to see the same level of enthusiasm among my colleagues.

In 2003, inspired by the aforementioned curriculum *Food Is Elementary* by Antonia Demas, PhD and that her daughter Ariel Demas continues, I began cooking with my patients. With her, I had understood how important it is to include all of our senses. I brought food that participants would touch, smell, chop, and taste, all whole plants. The result was promising with many smiling participants. I did not give exams or required comprehension tests. I trusted each would be grasping the material as it was pertinent to them and would use it accordingly. The process helped me learn how other people learn.

I also saw how powerful it is to share information in a group, where it comes not just from one person. A pharmacy student that was attending the classes said she had stopped consuming dairy the year before and her asthma disappeared. Another woman confessed through tears the harm she unknowingly was doing to her family with food. Some identified with her and responded with empathy and compassion. And then one person commented how much money she was saving, she did not quote statistics, she spoke of personal experience impressing those in attendance. The testimonies came unprovoked. The anecdotes contributed to the teachings like ingredients to the recipes. This collaboration could not be had any other way than in person.

One day I was invited to present to a group of gifted students of an elementary school nearby. I was warned that these were very advanced children. Apparently special enriching experiences are reserved for children already privileged. I brought my giant bowl and a box of produce, fresh vegetables and fruits, to make a rainbow salad. The teacher proudly welcomed me into an ample classroom wallpapered with large photos of lettuce, tomatoes, kiwi, cabbage, avocados, and apples. Each round table showcased cardboards with color graphs and more photos of food. I began my session holding up each ingredient for identification before the task. A chorus responded to the cabbage calling it avocado, the apple was called a tomato, the broccoli was tentatively called, a kiwi? and the cauliflower went unnamed. I felt shame for the teacher. It became obvious that cardboard renditions could never replace the tridimensional aromatic and delicious world of real food. On my way out carrying an empty box and my utensils, I crossed paths with a delivery person bringing them pizza for lunch.

On another occasion, I presented to a group of 25 students and cafeteria personnel at Penn State University about the planetary benefits of a diet based on plants. The chef, very attentive and courteous, complained that to accommodate vegetarians or vegans, he would need to prepare two menus or even more, which would be very wasteful of energy, time, and resources. I asked if he had thought of preparing "universal" soups, salads, even main dishes. He could use vegetable broth and leave separate the animal ingredients such as meats, cheese, and eggs. That way the students/clients could add them to taste. He opened his eyes and visibly relaxed exclaimed: Excellent, I never thought of it that way, of course I can do that! Sometimes we are not short on money, but on imagination.

GREENS ON A BUDGET—FROM 2003 TO 2023

In the year 2003, I purchased a 17-inch bowl, an electric portable stove, two large pots, a great stainless-steel pan, and a long handle spoon. Armed with a dozen chopping boards and twelve broad leaf chef knives I bought at a restaurant outlet store, I reserved the board room of the clinic where I volunteered. Right there we met monthly to prepare the same two recipes for two years: steel cut oats with quinoa and a gigantic rainbow salad.

No one complained that we made the same recipes, since we varied the fruits in the oats, and the grains, beans, vegetables, and leafy greens with which we made the salads. It was two years before I started adding rice dishes, soups, and desserts (always based on whole plants). New participants were always joining and month after month everyone felt more capable. The aroma of cinnamon and cilantro invited hallway passersby to curiously poke in their noses. The fact that only physicians and nurses stayed away surprises me still.

The reductionistic paradigm insists in prescribing drugs to suppress the symptoms of disease and the professionals who practice it are considered experts. Food is the thing of dietitians and women in the kitchen. Medicine has been dedicated to the creation of a know-it-all elite that must not be questioned. Arrogance and ignorance make a terrible combination.

If you visit the American Diabetes Association webpage, you will see that while their declared expressed mission is to prevent and cure diabetes and to improve the lives of people affected by the disease, they are mostly helping you live with diabetes. The monetary contributions of sponsoring pharmaceutical companies, such as Lily, NovoNordisk, and Sanofi, are over $53 million.

For the next fifteen years, I met to cook with interested groups in churches, schools, doctors' offices, community centers, orchards, and libraries, always with food, knives, chopping boards, reusable plates and forks for all, a giant bowl, a large pot, and a portable electric range.

As a result, some have made radical changes and occasionally some have seen the unexpected transformation of a family member. However, disinformation through the marketing and propaganda megaphone is everywhere and never ending, presenting a great challenge. The world is different and we have a new word that eloquently describes it: infoxication. It is difficult to maintain the new plant-based dietary pattern. Even people who have participated in the workshops for a long time are confused and tripped by the lies. It is reflected in their comments—what about chicken, what about protein, and what about eggs, and oil, what about supplements?

Too many people believe it is normal to suffer pain or have overweight, diabetes, or fatty liver. Generally, people commit to the sessions only once in a while—not enough. This information has to be studied, it has to be comprehended enough to be used comfortably. I know, because I have dedicated years to study and learn what I am sharing with you.

Every week, I carried two boxes filled with food in various stages of preparation to finish the dishes live with the participants. You could imagine how heavy these boxes were; plants are 98% water and a gallon of water weighs a little over eight pounds. I projected images on the wall. Printed information in Spanish and English, we read, discussed, and enjoyed the exquisite flavor of fresh plants. We learned step by step, one truth at a time.

Those who decided to stop milk, cheese, chicken, and oil reported with enthusiasm that in less than a month they no longer felt acid reflux, had more energy, and felt lighter. These rewards keep them going, but they must be lived. The experience of one does not translate into the action of someone else. This project shares true information that might be of interest to you, it could empower you to make better decisions for your own good, but you must live them.

FRESH2YOU

Occasionally, supermarket gift cards were donated to the clinic where I volunteer and I gave them to patients with a note that read only good for vegetables and fruits. Once a week the local food bank brought some produce that was quickly spent. Since I wrote for very few drug prescriptions, I began prescribing what patients should buy when they visited the market spelling out: cabbage, broccoli, cauliflower, mushrooms, beans, lentils, rice, blueberries, oats.

The program Fresh2You was established in 2017 and is still helping hundreds each year.

One morning in 2017, Roberta Cosentino came to the clinic accompanied by someone else from the food bank. She had heard of my food prescriptions and brought a smile born deep in a happy heart. We talked. That same afternoon with a generous grant from the Chester County food bank, the innovative Fresh2You program was born. Sixty families were enrolled in the first cycle with the help of our tireless Rosa, then there were one hundred, and today in 2025 the program is still standing. The truck comes loaded with overflow fresh harvest donated by local farms. It parks in town for two hours each week. Anyone can spend a few dollars and buy what they can, but for the families that present our medical food prescription, the food is free from May to November. Thank you!

ZOOM

During the pandemic, thanks to the support of the Guerrero family and Coalición Fortaleza Latina PA, we transferred our weekly sessions to Zoom. We celebrate these meetings free and in Spanish (when English speaking participants come in, we make the session bilingual) every Wednesday from 5:00 pm to 6:00 pm (eastern time). All are invited to join (including you). The code to participate is 879 7001 2349 and the password is plantas (lower case). We are ten to twenty people from all over the world. We dedicate the hour to learn and review the body and nutrition. We share comments, anecdotes, and recipes. I show educational slides and share strategies on how to make a transition to eating plant based. The participants reminisce about the food of their formative years, comment on the diet of their often-centenarian elders, participate in the joy, recognize the want, reflect on the injustices, remind each other that your struggle is my struggle, imagine a brighter future forged by all.

TWENTY FEET OF INTESTINE

This model of the intestine made to scale of colorful fabric helps us relate to the way our nutrients enter the body.

After the pandemic, I decided to meet again in person on Saturdays, this time in my home. Because we learn through all of our senses, I sewed a digestive system model to scale with multicolored remnant fabric. We arrange it on the table in relation to the skeleton, the heart, the liver, the stomach, pancreas, colon, and kidneys, placing emphasis on where the microbiota dwells.

We study food with the illustrated encyclopedia *Wellness Foods A to Z published by UC Berkeley*, which includes color photos of whole foods with their clear and objective nutrition labels. We incorporate informative facts in a web made of vegetables, grains, legumes, mushrooms, fruits, nuts, and seeds. We notice differences, similes, exceptions, affinities, and coincidences among the fruits, the leafy greens, the grains, the legumes, the nuts and other seeds. That way we improve our knowledge of whole food.

We compare lentils, garbanzos, kidney beans, soy beans, and peanuts with other beans or legumes; we study the nutritional transformation of a tender green bean to the dry mature bean. We see that all nuts have the same profile, except for chestnuts which are very low in fat. We make evident the nutritional resemblance between all leafy greens and other vegetables such as the squash and the chayote.

This exercise on food literacy feeds our innate curiosity. Using the encyclopedia as reference, the information is immediate.

PAPER AND PENCIL

Since 1970 every packaged food must have a label with basic information to get us started. To convert grams to calories and calories back to grams, we take out paper and pencil.

Each gram of carbohydrate or gram of protein renders the equivalent of 4 calories, each gram of fat is equivalent to 9. We calculate how many grams of protein, carbohydrate, and fat we should consume according to recommendations (10%, 80% and 10%, respectively).

For example: If we want to consume 10% of the 2,000 daily calories in protein form, what is the equivalent in grams? Well, 10% of 2,000 is 200 calories, divided by 4 calories per gram of protein = 50 grams of protein.

This exercise converts data into knowledge and when we know a material in depth, we can use it correctly to make decisions that benefit us. We need places and institutions that allow us to study this way, that help us keep and strengthen the truth.

CURIOSITY

The group is small and very few men have accepted the invitation, but onward we go.

Doctora, asks a student, why when a person with diabetes eats a tortilla or a piece of pineapple their sugar spikes right away, but the same does not happen if they eat a piece of chicken?

Silence... It is the time to see if they have been storing data or if they have true knowledge. I see them searching through notebooks and mental archives. They know there is an explanation. They've heard it in class. Where did they store it?

Without rushing to answer, I ask the student for more context. It so happens that her doctor has recommended she consume less carbohydrates to keep the sugar levels normal and noticed that avoiding the tortilla and eating chicken, the sugar did not rise. She was confused, since according to what she was learning, fat worsens diabetes which can be possibly reversed eating grains, vegetables, and fruits because they are low in fat. How will she reconcile this apparent incongruent information/experience?

It made my blood boil—because this (low plant, high chicken diet) is what many physicians are prescribing to manage diabetes. Why? Because they can achieve the desired sugar levels, but it does not revert insulin resistance or stop the disease.

The first part of the question has to do with insulin resistance. In diabetes, cells full of fat (in particular saturated fat) downregulate the insulin receptors (insulin resistance). In this condition, insulin cannot do its job of bringing glucose inside the cells, so any food rich in carbohydrates, such as tortillas or pineapple, will add more sugar to the blood.

The second part of the question has to do with the nutritional component of foods. The carbohydrates in the tortilla and the pineapple turn to glucose, while chicken with zero carbohydrates that could turn into glucose, won't elevate blood sugar in the short run.

This immediate way of learning, ties loose ends and lowers the risk of facts being blown away with the wind.

I will never forget what I learned from my first patients: the young woman who survived a pancreatitis after a car accident; the executive who stepped out of a seven-hour flight to suffer a pulmonary embolus; the mother with whom I sat for 48 hours as she came out of a hyperosmolar diabetic coma; the muscle fasciculations of a tall man with Lou Gehrig's disease. The names have faded, but I carry with me the beating of their hearts.

LATINO TRADITION AND CULTURE

The diet based on rice, beans, pumpkin, sweet potato, papaya, cauliflower, and other marvels of the soil is only one aspect of the resilience of our Latino life. Staying active, living in family communion, exuberant art expressions, dance and music, the sense of humor, and our innate leaning to celebrate, altogether create a positive exposome within which our genes have more opportunities to express health.

I grew up with the notion that in Puerto Rico people live a long time. My own great aunt Ninina, a healthy thin, active woman, would greet us in her little farm where she planted lettuce, strawberries, vegetables, and kept a few chickens. I marveled at her delicate needle work. Twelve years into their marriage, Pepe died of an anaphylactic shock at the dentist's chair; mourned him the rest of her life. She protected with layers of cream color paint a table he made with spent thread spools. At 96 she fractured a hip and died in bed at home with my parents.

Nani, her sister and my grandmother, claimed a full figure and a happy spirit. She stayed inside, loved her rocking chair, stepped to the patio for sun and to care for her plants, she taught me to crochet. Fritters and pastries were her weakness. Kept colorful candies and gummies in her bedroom closet and offered them to all her visitors. I was twelve years old when she died at 86 with diabetes and heart disease.

During my residency in Philadelphia, I worked with Latinos with cancer, arthritis, diabetes, gallstones, and liver cirrhosis. I don't ever remember diagnosing obesity (although there was), rarely depression (of course it existed), poverty was not mentioned, and of nutrition nobody spoke, since no one had studied it. In medicine, we disregard what we don't know. The epidemic of chronic diseases advanced without setting off an alarm. What we did experience was a wave of patented drugs for the management of their symptoms, the list grew each year.

In nearby area towns, I met people from Mexico, Guatemala, and various countries of South America. The face of chronic diseases defined. Elevated cholesterol was common, as was high blood pressure, fatty liver, and diabetes. I developed more curiosity for nutrition and sought to better inform myself about the past. I read *Harvest of Empire* by Juan González, a sweeping history of the Latino experience in the United States, and saw the systematic spread of contempt for frijoles and tortillas while the blind consumption of meat and cheese is enhanced and valued.

I realized that no matter if we come from the country or the city, we maintain the custom of eating rice and beans, plantains with yautía, nopales and avocado, tortillas, pomegranate, huisquil, mango and papaya. We flavor water with lemon, tamarind, and hibiscus. We favor milpa grown corn, bean, and pumpkin; our countryside grows banana, yautía, and cacao. In the kitchen, there are always pápalo greens, huauzontle and epazote. Our patios grow cilantro, purslane, tomatillo, chayote, and okra. We don't know what is to live without recao, malanga, and sweet peppers.

The North American lifestyle based on more of everything and everything bigger, has metastasized to every corner of the planet. The elders have memories and roots that keep them anchored, but our youth is growing without such souvenirs, to the tune of pizza, nuggets, fritters, cheese, a tablet, and influencers.

It is unfortunate that as professionals we have missed the mark. We believe the origin of this unbridled overweight stems from some personal psychological trauma, that it is dominated by practicing more willpower, that it could be cured with self-help courses, or that it might be conquered with diets, pills, and surgeries. We've seen such methods fail the majority of people.

I subscribe to the idea that the obesity epidemic was born in the novel and irresistible world we have created for our palate. We followed the trail of something sweet and fat to survive, but nature never offered them in factory combinations or in the obscene modern amounts. When you are reaching for a fritter, a glazed donut, a block of cheese, a piece of meat, remember that you are obeying your biological scheme, but in a fabricated environment.

The measure of obesity is alarming and also lucrative. To conquer the obesity epidemic, we will need to reinvent how to make money without destroying our health.

MY DREAM: LATIN AMERICA—ONE ENORMOUS BLUE ZONE

In the blue zones of the planet, we find many people who have reached the age of one hundred healthy and without disabilities. In parts of the world where everyone has access to the earth's harvest, where people participate of nature, where the wellness of the individual is paramount, and where one lives in harmony with the planet, people enjoy health for many years.

Los Díaz de la fotografía by Gloria Díaz de Puig (*The Diaz of photography*) is a beautiful project that captures the natural physiognomy of the Puerto Rican people from the nineteenth century to the middle of the twentieth. The book tells the history of photography through five generations dedicated to the profession. Until 1950, people displayed normal weight for their height; henceforth, their silhouettes began to widen. It is the same thing we see in countries around the world

Look for people more informed than you and plan to do something with them tomorrow. If today you find someone with more courage than you, do something together tomorrow. Make new friends and plan activities together. Today you feel trapped, tomorrow more liberated.

LOVE THE TRUTH

This is the story about a vain emperor that lives in the next county. One day, an unscrupulous tailor convinced him of making royal clothes that only those loyal to him could see. Word spread that the emperor had new clothes and once the invisible garments were finished, the swindler pretended to dress the king. The emperor did not want to admit that he himself could not see the clothes, but vanity won and he went out in public in his underwear. The people who remained loyal to the king pretended with him. More and more people had joined the farce when a child within the crowd yelled "the emperor is naked!" The whole troupe was agape, they had been fooled. Are you more interested in appearances than about the truth?

If everywhere you hear that we need protein from chicken and beef, that we must take magnesium supplements, that we benefit from the amino acids of colostrum, that this company's shakes are a marvel, that your bones need the calcium in milk, and that oil is gold, you may not be totally sure but since you don't want to seem ignorant, you join the farce and fervently defend the idea. How could you have fallen in the trap?

I am convinced that at our core we are thinking creatures, that we want to make sound decisions based in reality, that we respect scientific knowledge, that we are not fools. In order to create a world where truth prevails, we need to be curious and risk learning. Truths do not take care of themselves.

ABOUT FREEDOM

According to Timothy Snyder in *On Freedom*, to define freedom as simply being free of constraints or rules is to reduce freedom to its negative aspect. Bringing down a wall or letting one out of jail are requirements to be free, but by themselves do not create the conditions for freedom. We are free when we have access to a decent place to live, to protection, education, to health care, work, rest, and to the truth. To live free we must have access to true information with which to make decisions to improve our living conditions. A free state or government demands responsibility, exacts accountability, expects us to obey laws and assures the conditions that permit its citizens to prosper.

We have seen that many businessmen would prefer to deal free of boundaries, free of accountability, free of having to deal with only the truth. In our present historic moment, the extremely rich oligarchs want to eliminate all restriction and law in order to advance their machinations and amass more money than it is possible to imagine, without regard for who they rob or destroy. But businesses do not prosper on their own. They need partners, sponsors, and clients. Think of the mega industries to which you have subscribed because they provide you some short-term convenience or comfort, at what price?

USE YOUR POWER

We are the patrons, partners, and clients, the ones who buy and also produce clothes, food, books, furniture, medicine, cosmetics, kitchen utensils, cleaning products, the ones who give service to hospitals and hotels, in the end we are the community. We must demand accountability from all, advocate for the equal application of the law, protect scientific investigation free of financial ties, and not abandon independent free press plus we must impose limits on how much money a person can contribute to political campaigns. Together we have the power to form a free society in its full sense; one in which we can forge a healthy and prosperous future for us and everyone's families, one truth at a time.

LOOK IN THE MIRROR

1. The United States government subsidizes the cattle/dairy industry with money from the public and we follow its footsteps buying cases of meat and gallons of milk, when what we should really do is withdraw the undeserved subsidies. As individuals, right now, today, we can stop

buying meat and milk, especially when we know their true cost in personal and environmental health, and obtain better nutrition through plants. We have great power if we decide to use it on our behalf.

2. In 1985, it became legal to advertise prescription drugs directly to the public, who quickly obeyed and proceeded to request them from their physicians. If instead of responding with a knee jerk to suppress symptoms with a pill we acted to improve our diet and lifestyle, we could effectively resist this pharmaceutical propaganda.

3. When supplement companies promise the moon, we become their monthly subscribers. What if instead of filling ourselves with useless pills and powders, we nourished ourselves well and exercised? We know quite well that both of these actions favor us. What if we save the boatloads of money we spend on supplements and enroll in a nutrition course such as the one offered by The T Colin Campbell Center for Nutrition Studies?

4. When the industrial foods made their debut, we emptied our pockets in their coffers. We could stop this impoverishing hemorrhage by simply returning to our traditional roots buying earth products, cauliflower, rice, chickpeas, pumpkin, and blueberries. Love this body that they teach us to mistreat.

5. Unethical corporations deprive us of clean water, contaminate air and soil, and leave us in poorer health. If they amass millions, it is in part because we buy their products. Wouldn't you rather have the money you earn benefit you, your family, and your community?

6. It is true that we need universal coverage that responds to our health care needs and covers us throughout every stage of life, but we do not need medical insurance or large sums of money to eat more vegetables, tortillas, rice, beans, sweet potatoes, fruits, and fresh water. What we do need is to be educated with the truth.

The voice of special interests is spread by influencers, some are ignorant and others malicious, wanting to benefit from your innocence and perhaps laziness. They all use the same tactics: exaggeration, fear, and false promises. They pledge to free you from harm…for a price. Promise access to a special club…for a price. They alarm you with fictional diseases from which only they can protect you…all for a price. If you don't come armed with truths, you are caught in the net. Remember the parable of the emperor and his invisible clothes?

With this project we see that plants hold the treasure of health, that we don't need permission to eat from the earth's bounty, that the body knows what to do with it. The question I ask is, how will we discover this hidden treasure?

AMONG THE LETTUCES, ONE CABBAGE

For over twenty years I maintained a voluntary preceptor position in a family medicine residency. As one of the strategies for practicing medicine centered in lifestyle, I showed them how to do a virtual kitchen tour with their patients. This is when I met doctor Dan Chen. Years later he invited me to collaborate in a project to improve the health of his patients: the visits in Spanish followed a model of group education and cooking whole plants. His wife Charlene is medical director of that clinic. We all benefit from having support. For me, this was an extraordinary opportunity and a dream come true. Even today Dan continues building on his method and improving the health of his patients. His colleagues recognize him more each day and the community respects and loves him.

My colleague Dan Chen is an example to follow.

ANECDOTES THAT UPLIFT AND INSPIRE ME TO KEEP GOING

Food is a sensitive topic. Many people defend their gastronomic preferences in earnest. As a physician, I have interviewed thousands of people. They give me license (privilege) to enter physically and virtually into their homes and kitchens. I bear witness to human fortitude.

Judith and Ismael—at the vanguard

I don't remember who I met first, whether the brother or the sister, both of solid character with enviable imagination and creativity. As soon as they learned of plant-based nutrition, they applied their culinary knowledge at home and in the community. Judith quickly transformed the food at home and regularly presents our Zoom workshops novel dishes with various ingredients and spices. The week after we spoke of plant-based cheeses, she invented a cylindrical, elegant, simple, and delicious garbanzo cheese.

Recently she took her 14-year-old to a routine pediatric visit. As expected, the doctor asked the young man how much milk he drank. He proudly responded that his mom does not allow him to drink milk. It all broke loose. Judith was reprimanded. How could she be depriving him of the supreme source of calcium? In the midst of it, she said "we are plant based." The doctor immediately calmed down and the conversation resumed on common ground. We ought to ask, how and where did the doctor connect cow's milk with bone health?

Regarding brother Ismael, after a personal transformation, his generous involvement in the community especially with young ones leaves me in awe. Without ostentation or rewards and facing obstacles right and left, he keeps forging the way.

Antonio—returns to the soil, his home

When he arrived in the United States, his simple diet of rice, frijoles, tortillas, fruits, and vegetables radically changed. Within a short time, he was twenty pounds heavier and began to feel tired, with acid reflux, and his feet hurt. Now he is 40 years old, got diagnosed with high cholesterol, elevated triglycerides, and was told he has prediabetes. The doctor advised him to avoid fruits and tortillas because they turn to sugar. In our countries doctors, especially if they are men, are considered people of unquestionable authority, even those without nutrition knowledge. Sad but obedient, Antonio decided to follow the instructions. Then I received him in consultation.

Imagine his confusion when I wanted him to go back to his rice, fruit, and frijol. We had an hour-long conversation. Spoke of the elders in his country, of the harvest that nourished him in Guatemala, how priceless is our health. We spoke of diabetes and he asked me about eggs, shrimp, and the sticky myths relative to meat and milk. He began to recognize the storage of fat in his body as the result of what he had been eating since he came to this country. He started to understand the phenomenon we call insulin resistance and diabetes. He calmed down, seeing that he has agency and power, that eating healthy is within his reach, that his disease is not a mystery and yes reversible, that I would walk with him. However, the connection was still very tenuous. Out in the world the mirage will confuse him. We made a follow up appointment within a month.

We spoke again and told me he is coming along. Stopped eating meats. His pants feel more comfortable. Has no more acid reflux and his feet don't hurt. Timidly he mentioned that he stays hungry. In spite of our initial conversation, he heard from people that he should not eat rice and tortillas, but also fruit and sweet potatoes because they are too sweet. The power of propaganda! Our conversation was extended a few more minutes. I tried to help him understand that neither fruit nor tortilla cause diabetes, that the problem is the fat. I want to convince him that he needs carbohydrate rich food, precisely to generate energy, clean his cells of fat, and not stay hungry.

He has two sons and his wife four months pregnant has gestational diabetes. I suggest it would be much easier if the plant food pattern is followed as a family; they could enjoy delicious food together and for less money. Antonio agrees enthusiastically, but it takes two to tango and he has not had the opportunity to speak with his wife. We made another appointment.

Nely—her arthritis disappeared

The joints hurt her so much that she was forced to quit work. The rheumatologist had her on treatment with injections and pills, but could not control the pain. She entered my office holding her hands and dragging her feet. She reported drinking milk constantly to alleviate the acid stomach. She complemented her meals of meat, eggs, and fritters with some beans and vegetables. For a while we spoke about inflammation and food. Between feeling understanding and desperate, she agreed to make changes overnight and eat pure anti-inflammatory plants. Pain and swelling diminished in one week. The stiffness disappeared in less than one month, to the point that it allowed her to return to work. All of this without the side effects of medications. Two years later, Nely continued to enjoy total function free of pharmaceuticals. This is what sustains her, since she has no support from family or friends. That spark which provoked Nely's radical changes happens very rarely.

William—(tentatively) reverted his diabetes in just three months

Height: 5'10" / Weight: 165 lb / Weight at age 20 years: 120 lb.
At 32, He arrived with a glycosylated hemoglobin of 13% (normal is < 5.7%). He consumed fried meat and pan dulce (a typical sweet pastry) every day. In just one session, he understood what he had to do. With help from his wife, the entire family changed their eating pattern to whole plants they could pick up at the Fresh2You food bank truck. Taking the oral medication metformin, in three weeks his blood sugar came down from 300mg/dL to 140mg/dL. After two more weeks following the plant diet without oil, he reached blood sugar levels of 100 to 114. In two more months, he had lost 30 pounds and his HbA1c of 13% had come down to 6.7%. What a success! Naturally, he rested on his laurels and one year later his HbA1c had climbed to 7.3%. Recognizing health does not take care of itself, he fortunately got back on track. These teachings are very valuable, as long as he does not veer too far from the path.

Carmen—searched and searched until she found

She tirelessly advocates for the community, where everybody knows her, but she had obesity, inflammation, suffered from acid reflux, had a sundry of digestive problems, and a chronic debilitating infection. Not yet 50, she felt spent. Depended on drugs and energy drinks to get through the day. By chance, she came to one of my workshops and what she heard shook her to the core. They say that a word to the wise is enough. That same day she threw herself fully to the task of cleaning her kitchen and her diet. The daughters alarmed watched as she threw out all meat, milk, eggs, cheese, and junk food. The rest is history.

Her body quickly responded, returning to the health she feared lost. To her surprise, she had recovered with the same nutrition she had as a child. Her activism changed course. However, don't misunderstand me, I don't want to give the impression that a radical change is so easy, the information has to activate the person in a way that escapes logic and in a way for which there is no formula. Today she is 60 pounds lighter, feels how every cell of her body has revived, and fervently protects them. Her greatest wish is to see the young grow healthy. The 4-year-old grandchild is forming very close to her, delighting in the food and adventures of his abuelita. There is no greater reward. Her dedication has not diminished. She knows what can be accomplished with the effort of many more like her. She carries her health like a shield motivating those who have witnessed her transformation and want to follow her.

Rosalba—example of tenacity

Height: 5'1" / Weight: 185 lb.
At 53 she was diagnosed with fatty liver and prediabetes, but this started much earlier. Twenty-five years ago, when the family arrived in the United States, they quickly got used to pancakes, eggs and sausage breakfasts, and what she calls street food. In the blink of an eye Rosalba gained 60 pounds. Had surgery for gall bladder stones and very elevated levels of cholesterol and triglycerides. Worried, she started to walk, to jog, and to eat better. This presented a challenge since her family did not join her. Attending the workshops, she learned to make recipes that pleased them, with ingredients like tofu, soy curls, oat milk, cabbages,

and more spices. Friends suggested she use pills to lose weight and felt tempted, but after considering and investigating, she decided to keep up with the food.

Not much later, she shared with the group that the ultrasound no longer showed a fatty liver, her glycosylated hemoglobin and cholesterol levels were normal, and she had lost 14 pounds. Her doctor rewarded her by discharging her as a patient. Apparently support does not come in a pill. She is not yet standing on firm and safe ground, she has barely walked from the edge of the precipice. Holidays approach when traditions call for sharing with friends and family the confections that made her develop the conditions she is trying to leave behind. Today more than ever, our group will be her anchor.

Lety—for many years believed she was Wonder Woman

Height: 5'3" / Weight: 130 lb || Weight at age 20: 125 lb.
Had a surgical menopause without hormone replacement and reached a 25-pounds overweight. By her own account, her early diet was wretched, but felt invincible, since she ran marathons. She tried to reduce carbohydrates, but she loved bagged snacks and depended on ready-made food, for she did not cook. She was surprised that although she exercised and drank milk by the gallon, her bones had lost calcium (mineral). Today she is 66 and hasn't used tobacco in many years.

Alarmed by her elevated cholesterol she tried to educate herself. This time she based her changes on scientific evidence. Her first transition was to plants, mainly processed. Little by little, cheese disappeared from her menu, then reduced flours and added sugars plus the olive oil. The nuts and avocado portions grew smaller. When she stumbled upon our Saturday group, she became one of the most assiduous participants, famous for bringing refreshing one ingredient dishes: grapes, roasted Brussel's sprouts, baked tofu, cubed melon, oats sweetened with dates, or mandarin wedges. Now, she offers rice with lentils and vegetables, seasoned tofu scrambler and other delightful greens. The progress is evident!

She began this lifestyle quietly, at her rhythm and without hurry or embellishment, with many questions and always with her eye on the prize: more health. Reading labels to assure she's buying packaged foods that conform to the basic requirements : no added oil, no added sugars, and low in sodium. She is irritated by the insensitivity of people that won't move a finger to make dishes universal, and feels pushed in a culinary corner. Her challenge is to learn how to request without being upset. Very soon after her nutritional changes, she began enjoying levels of health like never in her life. Anyone who doesn't see it, is afraid to face their own decisions.

Maritza and Olga

The Colombian sisters are a marvelous and beautiful complement to each other. They entered plant-based nutrition many years ago through the door that spelled health and adopted this lifestyle without hesitation. Oriented to the family, they share their delicious creations without imposition or proselytism. As they continue to connect this lifestyle with themes of animals, social justice, and the environment, their dedication to constantly learn is contagious.

María—out of the danger zone in just six months

Age: 45 / Height: 5'1" / Weight: 135 lb / Weight at age 20: 120 lb
María tells us that when she began attending the virtual workshops, she had a diagnosis of prediabetes and fatty liver. She felt fatigued and satiated her hunger anxiety with pastries and fast food. In her kitchen there were always meats, chicken, and eggs. When a friend told her of the workshops, she began attending week after week; sits in front of the camera in a classroom, attentive and always with a smile. Occasionally contributes with on point positive comments.

One afternoon, she took the mic to announce that her lab results, which had been quietly improving, were now all normal. From coast to coast we celebrated with María. According to her, by eating less meat, chicken, and eggs, she is recovering health and has lost five pounds. She has conquered the family with a signature salad of black beans, nopales, cilantro, and lemon. At her home we can now find more vegetables, asparagus, and broccoli. Her husband is a valued support, the 4-year-old grandchild is the most fortunate, her pregnant daughter does not have it so easy. She now calms her anxiety with a smoothie of pineapple, cucumber, and celery. With no complaints or excuses, she is an example to her family and to her community. María knows she has a long stretch to walk and is persistent as she redoubles her efforts.

Álvaro—recovered with the food of his country's elders

Age: 47 / Height: 5'11" / Weight: 284 lb / BMI: 39 / B/P: 138/94 / SGOT: 373 / SGPT: 133 / HbA1c: 6.7%
Álvaro came to the United States 20 years ago and quickly adapted to eating the American style. Reached his heaviest of 297 pounds with a daily diet of meats and fat rich food. Lately had not been feeling well and on his own changed his breakfasts to green smoothies. When I met him, we spoke of what elders eat in his country and he reconnected with a fiber of his culture that he had semi forgotten. Lives with his wife and children, now 27, 20, and 14. At our second visit, he reported with satisfaction that the youngest, who used to drink oodles of milk, is now on a dilution of cow's milk and almond milk of 50:50 without complaining. He gets emotional telling me that his wife is his greatest support and that he has already lost eight pounds. In my agenda, I have written to call him every few weeks. This is a very small support, which I pray he eventually can find among his own.

Ángel—after a scary hypertensive crisis he made a transition

About ten years ago, Ángel had an episode of atrial fibrillation that threatened his life. He was discharged from the hospital on medications to control his hypertension, without a word on the lifestyle changes he could incorporate to treat the overweight and triglycerides which were very elevated. When we met, of course we spoke of his nutrition. He took the reins temporarily, made a transition to eating whole plants, lost weight, and normalized his laboratory results. However, he lives alone and depends on the food that others cook. When he stays with his daughter in law, he eats super well with lots of vegetables and fruits. But when he lives with his work buddies, his diet reverts to fried meats. He had a couple of ugly setbacks when he gained weight, his blood where their diet of mostly plants has been replaced with hamburgers and French fries. More than a figure, people lose health.

In the Mexican population, 0.12% (1 in 1,000) is one hundred years or older. If we compare statistics, in the United States only 0.02% (1 in 5,000) is centenarian; in Puerto Rico, it is 0.03% (1 in 3,000), mostly women. Following the western diet, we allow health to escape like water down the drain. If we were to return to the traditions of our countries, we could cultivate an enormous blue zone comprising all of Latin America.

STORIES FROM THE COMMUNITY

Tell me who you are and I will guess the company you keep. I would like to report that truth dethrones myths just because, but it does not seem to be so easy. Myths are widely spread false beliefs or ideas that have found a footing in the psyche of people. We need the truth to replace them, but that is hardly enough.
With the exception of hermits, we all long to be part of a community, to belong. So much so that for the sake of being accepted, we yield and let go of many of our needs, because the other side of the coin is to be rejected.

There are five recognized categories of our personalities that benefit us if we strive to develop them more or less equitably: openness to new experiences, conscientious responsible behavior; extraversion/sociability, agreeable/pleasing disposition, anxious/neurotic tendencies given to depression.

If we seek the company of people who show compassion, who calmly navigate new situations, who welcome different voices, who take turns, who are concerned with the repercussions of their actions, we join in and tend to grow in the less developed areas of our personalities. The group is obviously healthier when its members expect growth from each other.

By contrast, a group which insists on submitting to ideologies or demands that you relax your scruples, that isolates you from others, a group that is exacting, or is given to catastrophizing will place barriers before your growth. Watch out with whom you associate.

pressure and triglycerides spiked, and developed hypothyroidism. Given his work schedule, the visits are sporadic. Remembering how well he feels when he is on a plant diet, he himself gets back on track. As time goes by, he is more capable and his good habits last longer. This lifestyle must grow to become a second skin.

Norvan—ate cheese for a year and his face blossomed with acne

At 29 he had to interrupt his studies and place his aspirations in limbo, to come work in the United States. Immediately he was initiated on pizza and fast food. Horrified, in just a few months he saw his face transformed to a grate of red pimples dotted with white pus. One day he came to the clinic curious about what we could do to help him. In no more than ten minutes we suspected that the daily cheese he had never consumed in his country was to blame. He left with facts and hope, plus a lotion for temporary use. Work took him outside the area and we lost touch for a year. One morning he showed up to let me know that in just three months back on the diet without cheese of his country, the acne had disappeared and his skin was new.

Luis—lowered his HbA1c of 12.8% to 6.7% in two months eating plants - all it took was an enlightening consultation.

Whenever I lose strength or feel weakened, I meet someone like Luis who assumes responsibility and recovers health without any effort on my part. I live eternally grateful to people like him.

REFLECTIONS

Too many of us have fallen onto a diet short on fiber with foods low in nutrition and high in animal protein and fat. This persistent insult causes inflammation, weight gain, and insulin resistance. Elevated levels of cholesterol and triglycerides pave the way for the silent development of arterial plaques. The average sugar in the blood of those diagnosed with diabetes is between 300mg/dL and 400mg/dL (normal levels are 70 to 120). All this punishes a body under pressure that enjoys no days of vacation from disease.

LOGIC AND DISSONANCE

Logic validates reason, logic is calm; dissonance combines elements that do not harmonize, dissonance is tense.

A healthy diet should harmonize with its results. In other words, if we follow an optimal diet, we are supposed to enjoy good health. Having to take medications to treat high blood pressure, cholesterol, and inflammation is dissonant. Feel the tension?

WE CAN HELP PAVE THE WAY

The path toward health is filled with obstacles, we can all help remove them. Owners of taquerías and small restaurants where the workers eat their traditional food, could make money selling tamales, sopes, tacos, rice and beans, mofongo, pasteles, arepas, sushi, spaghetti, even pizza using healthy whole plant ingredients. I know food establishments that are already popular and attract a steady clientele. They could be delicious shining agents of public health instead of contributing to disease in the community. Beloved establishments have the power to be promoters of freedom instead of securing the chains. If they only risked learning about nutrition, remembered how our elders ate, and maintained the tradition in practice! I dare to bet that their clientele would grow.

If you are a school teacher, any discipline lends itself to educate the students about the nutritional needs of the body and to immunize them against junk food. Imagine in Art class designing a mosaic with dry grains, beans, nuts, and seeds; or a high school Math class promoting literacy calculating the percentage of calories and daily grams of protein and fat that we need. Geography students could compare the land necessary for plant-based agriculture versus animal agriculture. It would be exciting to learn, for example, that an acre of land can grow 40,000 pounds of beans, while the same acre of land would yield only 400 pounds of beef. Music is universal and has no boundaries. Likewise, the news of the day, fables, kitchen recipes, these all lend themselves to study the written word and can expose us to true and practical information. The athletic output depends on excellent nutrition, students can carry out experiments and see the short-term results. Put away the tablet, turn off the screen. Learn the physical properties of the food nourishes us, we all have a body.

Avoid bringing addictive food into the house. If the institution where you work operates in ignorance, raise your voice and share true facts. Don't be intimidated or lose courage before the powerful. It's up to each of us to protect and defend the truth.

If your source of information is media on the phone, do you have it tuned to some commercial propaganda or to a channel that exposes you to influencers? It takes more time to read the written word, because you must focus on the page; audio messages fill your head without giving you the chance to pause to reflect. Audio messages entertain you while you work. Later you don't remember where you heard the lies.

Data from independent investigations do not stream like the opinions flowing off the web. Do you remember who told you to buy bottled water? Where did you get the idea that you need more protein? What reason did they give you? Perhaps you heard it on a commercial ad? It is very likely that the idea came into your life while you were unaware or while imitating someone that had influence on you.

The path to true information presents obstacles. There are 64 million Latinos in the United States, each of us came in through a different door. Most of us arrived to this land of plenty in relatively good health. When facing basic necessities you don't have much time to think of anything but work. We stumble onto a diet of fried meats, sweet pastries or pan dulce, easy eggs, and energy drinks.

The fast food of most Latinos is not the typical McDonald's or even the pizza of places where they might work and not all fall in love with cheese. The fast food to which many Latinos fall prey is pan dulce (sweet bread ingredients: flour, evaporated milk, white sugar, melted butter, eggs), carnitas (seasoned pork simmered in lard until it can be shredded and roasted to a crisp), pork chops, ribs, arepas and other fritters, pinchos (skewered roasted meats), fried chicken, rice with cured pork, and chicharrón (pork rinds). This and a sweet drink is the choice food. The obesity epidemic is not a mystery.

MORE ANECDOTES

The stories that follow collect some common characteristics that in my experience block success. Study each one. Pay attention to the stated problem, the laboratory results, and the diet. Take into account the height and weight of when the person was 20 years old (this is the best frame of reference for someone's ideal weight). Note the surprising weight gain that brings almost all of them into the range of obesity. We are talking of young people, working for who knows how many months paying no attention to their wellbeing, before being diagnosed with a disease, because work is paramount and rightly so. The names used throughout are not real.

Selma—complains of obesity

Age: 37 years / Height: 4'11" / Weight: 160 lb / Weight at age 20: 120 lb
Diet: Her refrigerator is filled with chicken, beef, cheese, eggs, butter, cream, sausages, soda, fruits, and vegetables; she replaces half a gallon of oil every month. In her first attempt to improve her nutrition, she followed it as a manner of diet, reducing her portions and concentrating on salads, but she was hungry and stopped. This is a common mistake. It took us several conversations for her to understand that transitioning to plants is a new way of eating and not a restrictive diet, but instead one that provides a greater selection of foods. She listens to so many comments that stop her. Finally, although hesitantly, she agreed to eat more sweet potatoes, frijoles, fruits, and vegetables. She was amazed! The stored pounds began to disappear, but these success steps do not endure by themselves. We doctors still do not understand the importance of creating an environment of solid support that could be maintained for at least a year, during which the person can navigate successfully through a cycle of anniversaries and key holidays, such as birthdays, family reunions, religious celebrations, and the like.

Carlos

Height: 5'3" / Weight: 210 pounds / HbA1c: 11.3% / Chol: 246mg/dL / Trig: 1,000
For this single dad with three young children, the primary focus of his life is work. He maintains a diet of fried meat, eggs, rice, and noodles.

Every two days, he eats beans, cauliflower, vegetables, and fruits. For stamina he depends on energy drinks. He has always been overweight. Received a diagnosis of diabetes three years ago and ignored it. Was hospitalized for pancreatitis some months ago and as soon as he recovered enough, he went back to work.

On the first day we spoke, he radically changed his diet, but I have learned that without social support these changes do not last. In the meantime, he got sick with a bad cold. There are no words to express his stress at the loss of income. In a week and barely able to stand, he went back to work. Nonetheless, while in the middle of all this, he maintained his new diet, was able to stop injections of insulin, and his blood sugar levels remained below 130. He is also proudly adapting his children. He says it is a bit hard, but is committed to this food pattern for life. We hope. His C-peptide level is 4.8ng/ml which demonstrates that he produces enough insulin. His triglycerides of 2,000 put his pancreas at great risk of another pancreatitis. I try to speak with him weekly to support him for the development of lasting habits. How many patients can a physician see like this, one by one?

Ingrid—worried about obesity

Height: 5’2” / Weight: 190 lb / Weight at 20 years: 120 lb / acid stomach
She eliminated pan dulce (sweet bread pastry), increased the salads, and in one month did not have any more acid reflux. She must go on by herself without current community support.

Catalina—puzzled and baffled with her deterioration

She grew up svelte, in constant movement, and with lots of energy, eating vegetables, frijoles, and everything that grows naturally in her country. At age 27 she risked crossing the frontier to escape violence. In a couple of decades her body stored 100 pounds of fat and visited the doctors with hypertension, elevated cholesterol, diabetes, psoriasis, and a very painful arthritis. On the day I saw her, she was rejecting her own body and blaming herself, for who knows what.

Doctors had been filling her with drugs to quiet her immune system on fire. Their plan was to keep giving her drugs, watch for side effects, and address them with more drugs. When we suppress the immune system, we suppress all our defense responses, the damaging chronic inflammation and the acute inflammation that could potentially save your life. She began to have frequent urinary infections. Her weight continued to go up. The frustrated rheumatologist gave her an ultimatum: lose forty pounds or I won't see you anymore. They say that the client is always right, but apparently the patient is to blame when she doesn't behave according to the medical paradigm or plan. Her doctor, rather than showing curiosity for why her tools didn't work or were inadequate; instead of learning the relationship between chronic inflammation and the patient's dietary pattern, preferred to eliminate her from her practice. Out of sight, out of mind—I prefer the literal translation from Spanish: eyes that don't see, heart that won't feel.

Catalina finds it impossible to think that the coveted lifestyle of this country supreme could be hurting her. Although she has not stopped eating frijoles altogether, she considers them food for poor people. Insists that she is eating well, because she is imitating the wealthy North Americans that prefer chicken instead of pork, fry in olive oil instead of lard, and drink juice instead of soda. Nevertheless, she is sick, so what more evidence does she need?

If she looked up, she would see that 60% of the country is in the same boat, inflamed and about to capsize with her. Will she change course? How can we help?

Julia—returns from her cardiologist

Julia suffers with chest pains and gets winded during exercise. She insistently explains how she is active, walks, swims, and goes to the gym. Recognizes that she has obesity, but cannot comprehend why she has diabetes, elevated cholesterol and triglycerides of 900mg/dL (normal is less than 100). Lives anxious because her father died of an infarct (heart attack) after years with diabetes, high blood pressure, and very elevated triglycerides.

Based on an echocardiogram, a cardiac stress test, and a coronary calcium score, her cardiologist concluded that "the patient is a very low risk for an infarct." Was I reading someone else's report? Julia has obesity, diabetes, very high lipids, family history, and suspicious symptoms.

Her kitchen reveals: garlic, onions, ripe plantains, rice, pasta, beans, oats, whole cow's milk with enzymes (the family is lactose intolerant), ham, sausages, cheese, two dozen eggs, butter, yogurt, beef and pork meat, chicken breasts, fish, shrimp, and a gallon of oil.

The cause of her obesity, high cholesterol, and diabetes is clearly exposed in her kitchen, but she is convinced it is all hereditary. Insists that removing the chicken skin, cooking meat on the skillet, reducing the number of tortillas, not drinking soda, and changing to olive oil ought to be enough. She clings to these cosmetic changes even when they have yielded no results. The real reason that keeps her from making consequential changes is that every week she goes out with a group of friends to places where the menu consists of fried meat, eggs, and pastries. She confesses that adopting more changes would jeopardize her social life. Unless she finds a novel way to feel socially connected and supported, her disease will march on.

Juan—added himself to the diabetes epidemic

At 33 and following a diet of meat and beer, Juan transitioned from obese to diabetic. He refuses my recommendation to eat plants, because according to him that is extreme. Besides, he knows that men need meat. He cannot accept that he has diabetes. He fulfills his work agenda which requires him to stand very long hours in order to send help to his family, which he hasn't seen in several years. His dreams are by now a faint memory. He is not interested in learning to cook healthy. He already threw the towel. Shows up to his visits as a responsibility. His blood pressure is high. Doctors are watching his kidney function.

Diana—at 35 showed up with diabetes

Age: 35 / Height: 5'5" / Weight: 190 lb / Weight at age 20: 130 lb
Diana remembers that as a child was poor but healthy and always with energy enough to play. She ate simple beans, tortillas with vegetables, and fruits whenever there were any. Arriving in this country she quickly fell into the clutches of the charmed food: fried meat, cheese on everything, pastries, and energy drinks. Admits she feels powerless in the addiction, even though confronting it could restore her health.

We live in an environment filled with contradictions that we do not question. I witness this lethal lack of curiosity and ask myself, what makes anyone wake up? But there is no collective waking up; we wake up one by one. Perhaps that is why mass efforts are so poorly rewarded. In an environment of disinformation where factual education is lacking, a person trying to make good decisions is walking a mine field, sometimes you make it and sometimes you don't.

Sergio—was in shock when he learned that his prediabetes was now diabetes. Eats meat accompanied with yogurt, eggs, and cheese; and buys olive oil by the gallon.

Kelly—follows information on the web and does not understand why she has diabetes.

Age: 30 / Height: 5' / Weight: 164 lb / Blood sugar: 430mg/dL / HbA1c: 8.9%
Kelly's diet consists of whole milk, chicken, eggs, sausages, ham, refreshments, and she buys oil in liter containers.

Gerry—says her diabetes type 2 is hereditary because her parents have it. Her very restricted diet consists of pan dulce, fruits, milk and cookies.—this is what she feels her money can buy.

Elvin—young Elvin yawns during the consult, because he has sleep apnea. He lives with two brothers with diabetes. Their diet consists of meats, fritters, eggs, and beer.

Juan—migrated healthy and in two years developed diabetes

His daily fare is fried meat, chicken, milk, eggs, cheese, tortilla, noodles, avocado, some vegetables, cauliflower, cabbage, and boxed cereal.

Fidela—gained 100 pounds without even noticing it.

Fidela is attacked by ignorant people who make hurtful comments—you were so pretty, why did you let yourself go?—if people knew anything about the obesity epidemic; that it has a strong environmental scientific basis. If she knew that it is not about restrictive diets or will power, she could deflect ignorant comments. She is young and is here by herself, easy target for those who have nothing constructive to contribute. They themselves are overweight and diagnosed with the popular diseases, powerless, yet find it comforting to put her down, to criticize. Yes it requires effort and yes it has to do with her food, but most of all it requires true information.

It is not easy to come out of a disinformation hole. At first, it will require a good dose of energy and it will certainly be easier with someone for company. However, do not feel shame or believe it is your fault. Be curious, check out reliable sources like the T.Colin Campbell Center for Nutrition Studies. Study, become more conscious, draft a plan. If your strategies are weak, you will be at the mercy of last-minute decisions. You must learn to prepare food in advance for when you don't have time or you are back from work tired and hungry. You can and should eat healthfully without sacrificing taste or the family's preferences.

The majority of the people I see stock their kitchen with convenience food. They are neither idle nor lazy. Quite the opposite, they are industrious people, many are working two or more jobs, sacrificing for what they need, the literal translation of the phrase in Spanish is "turning tripe into heart." They all reached an obese state, inadvertently.

References:
The Pleasure Trap, Doug Lisle PhD,
Hundreds of recipes at the *Center for Nutrition Studies T. Colin Campbell*,
Adriana Cortés MD, *Bien Plantado and the TEN program*

The influencers know how to work today's means of communication to spread their fraudulent messages. They borrow scientific terms (leaky gut, alkaline diet, deficiency of "X" or "Y," essential nutrients, etc.) but they do not follow scientific rules. In other words, they do not support their allegations with peer reviewed studies or vetted facts. For the person who suspects that the authorities are deceiving them, the influencer's rhetoric tends to confirm their suspicions. Does it matter that you are being used to sell a worthless product for financial gain? You fall for a plan meant to make you feel part of an elite that shares something in common, something special.

When health professionals fail us, what recourse do we have? The following short anecdotes include data from the Virtual Kitchen Tour.

Carla—receives assistance from the government.

Age: 29 /Height: 5' /Weight: 160 lb /Prediabetes with C-peptide of 1.9
The daughter's pediatrician recommends that she give her cow's milk three times a day. Women Infants and Children (WIC) supplies this milk plus cheese and eggs. Both mother and daughter suffer from constipation and the two-year-old has iron deficiency anemia. Did you know that casein (milk protein) is associated with slow gastrointestinal transit (constipation) and intestinal inflammation with microscopic blood loss (anemia)?

Petra—her dietitian prescribes her to eat chicken and eggs to control her diabetes.

Age: 42 /Height: 5' /Weight: 140 lb /Diabetes, HbA1c: 11% with C-peptide of 2.1
The kitchen reveals one half gallon of milk and one gallon of oil, 18 eggs, butter, cheese, 2 chickens, ground beef, and salmon. The prenatal

dietitian instructed her to eat meat to control her sugar. It is not working, but it does not matter because the dietitian is a professional that has been seeing her for a while. I am new to her care and appear to be contradicting her suggestions. Of course, how is she supposed to believe me on our first visit? She needs time to study, to learn about the body and about its food. That is the only way she might be able to make decisions based on truth, but who has the time?

Elena—her husband has gout, they both have diabetes, and fatty liver.

Age: 38 /Height: 5' /Weight: 239 lb
The couple consumes chicken, milk, and eggs with butter because they like the food too much, end of story. They own a restaurant. I can spot the conflict of interests, how to address it?

José—lives bound to the idea that meat, eggs, milk, and cheese are essential.

Age: 35 /Height: 5'7" /Weight: 300 lb.
José insists that he is not sick, because he feels fine. Besides he is sure that diabetes is hereditary and has nothing to do with the food. That's what people say plus he hears it from the internet. So he will continue to consume everything fried in lard (pork fat) and drink alcohol every day. Okay.

Miguel—argues that he needs to eat more protein, he is a man, everyone knows this.

Age: 38 /He is 20 pounds overweight, out of shape and with very elevated cholesterol.
In his home country, Miguel used to play soccer, but since he came to the U.S. works long hours in a pizza place. He ponders about going to the gym, but has no time. During the very short visit we had, I suggested he consider altering his diet of frijoles, chicken and quesadilla for frijoles, rice

and vegetables, but a friend told him "you need protein" and his friend's argument won. Miguel knows nothing about his body requirements, does not have time to learn, and fills himself with propaganda.

Sofía—the 10 pounds at the university

When Sofía complained that she had gained 10 pounds, her classmates laughed and said it is normal. Away from home and desperate she dove into the internet for how to lose weight. She found someone on YouTube convincing her that "carbohydrates make you fat" and of course offering a wonderful supplement that promised to lose pounds. Pum! She swallowed it. Right away she eliminated fruits, oats, frijoles, and sweet potatoes from her plate and sent away for the wonderful product. (If people were this quick to eliminate what is truly hurting them!) Embarked on the first of many diets and plans, none have had lasting results. If she only knew that fruits, oats, frijoles and sweet potatoes were never the problem, but that instead the problem was the fast food, the regular bags of chips, and her daily pastries. Where will she go for correct information about nutrition? Will she continue to surf the web for another guru?

Rosario—and the poisonous beans

Rosario heard a nutritionist on the web saying that beans have poisonous lectins therefore we should not consume them. This information, tragically incomplete about lectins and beans/frijoles, has invaded the internet. Why aren't you already dead eating beans? She remembers eating beans all her life with no problem, but now is paralyzed with fear and decides, no more beans! She denied a lifetime of experiences simply because she believed in someone that wants to become famous ruining the reputation of beans. (Check out the section on lectins)

GLOSSARY

The following glossary will help you interpret the previous stories.

Height of 1.52cm is equivalent to 5 feet (60 inches).
Weight of 50kg is equivalent to 110 pounds.
Body Mass Index (BMI = kg/m2) serves to compare the same person at different ages.
Normal total cholesterol level is less than 150mg/dL and triglycerides around 100mg/dL.
Blood sugar level should be maintained between 70mg/dL and 120mg/dL.
Glycosylated hemoglobin HbA1c is the percentage of hemoglobin molecules coated by glucose;
The normal level of HbA1c < 5.7%.
The level of C-peptide in blood represents the amount of insulin being produced by the pancreas.
The normal range of C-peptide is between 0.5 and 2.0 nanograms per milliliter (ng/mL).
Whole cow's milk is also known as full fat milk. This milk is distinguished in the U.S markets by its red top. Whole milk has the original amount of fat (45% of its calories are fat).
The industry sells this milk as 3.5% fat; but this designation is by weight, and sounds less fatty. Similarly, the milk 2% and 1% also represent fat content by weight, fat is diluted in more water.

LIVING THE CHANGE

The behavior of people ruled by ideologies or who live in governments ruled by dictators is regimented and preordained. However, the behavior of human living free is not dictated, it is unpredictable.

When people are educated with the truth, they can make their own decisions consciously with integrity; people educated with the truth are free to act according to their values.

In 2003, in order to advance our body literacy and food literacy based on plants, I decided to meet with groups of folks; first in a clinic and then in churches, in education centers, in the community, libraries, public gardens, in the house of a neighbor, and most recently in my home.

This collage of tables represents over a year of Saturdays that we call the Breakfast Workshop where we learn and enjoy in community. Visitors come and go joining the assiduous group that is intent on improving their knowledge based on true facts to share with the world.

I have learned that the instruction or advice from others takes a person only so far. If we really want to see changes in habits and behaviors, we need to study with people more prepared than we are and practice what we learn, preferably in communities of two or more. An organic community group is ideal. Otherwise, medical group visits like the ones carried by my colleague Dan Chen, support practical changes in people's patterns of nutrition.

IT IS NOT EASY

This may seem obvious, but to stop the use of oil you first must want to stop using it and second, you should know why. If you do not know the answer to these questions, it will be near impossible to keep to the new ways and any strong breeze will throw you off course.

To successfully leave the comfort zone, we must aspire to something different or feel that the current routine no longer works. Because even when we profess change, we vacillate. We humans are experts at fabricating a thousand and one excuses to avoid full commitment to the task. We make a feeble attempt, misinterpret the short-term results and turn around. I've heard people say that their nails became fragile or their skin turned dry. That is a classic excuse to go back to using oil. We want to stop drinking alcohol or want to exercise, but we put obstacles in our way. Obstacles come in the form of other people, false narratives, this or that guru, or convenient propaganda.

Recently, some of my patients are quoting a "Nutrillermo" on the internet, this character has two million followers. From him, they are "learning" that oats are for pigs, that pork meat is healthy, that we should use lard, that people should stop eating sweet potatoes and fruits because they are too sweet. All the while he peddles his line of products. What is the fascination with someone who spreads these false narratives and quotes his mentor saying "a millionaire knows how to make money, a billionaire knows how to make others make money for him." He is talking about you, dear reader, you are making money for the billionaire.

We don't need to go far to see economic interests paying for the medical information that my patients receive, which instead of helping them improve their bad habits help them opt for pills and shots. This is how I found one morning how many of my patients with diabetes at the clinic had been given a prescription for the injectable GLP-1 drug Ozempic. Once hooked on the drug, incentives to improve the diet move to last place. In other words, accepting the drug replaces the possibility of cure.

THE SATURDAY BREAKFASTS

Once we were out of danger relative of the pandemic, eight people signed up to meet on Saturdays in person with the purpose of learning about the body, about the optimal diet, and to celebrate in community. Of course, we also enjoyed plant-based snacks.

Each person attended without fail and the group grew strong. They spoke with enthusiasm one morning on how to share what they were learning. Soon the frustration was palpable. They complained that other people don't understand, that they are always trying to explain themselves. It is funny how we can walk a lifetime in the conventional way, but when we lose the blindfold we also lose patience with others who now must show patience with us. Ha! We want to see rapid change forgetting that up until a few months we were there with everyone else. We see the same impatience with people when they quit smoking or stop drinking. The main important task ahead is for us to foster compassion as we create community.

Some participants have decided to stop using oil and are irritated that everything in the restaurants comes with oil. So they tell the waiter that oil is too processed and could contribute to the clogging of arteries. At the moment, this answer skips many steps plus offers no alternative to the waiter. We would need to say that fat impairs the production of nitric oxide by the endothelium, which causes the arteries to turn temporarily rigid and makes blood sticky. If we were to complete the concept, we would need to add that the liver uses the saturated fat in the oil to make excess cholesterol, which when oxidized and ingested by the macrophages forms plaques in the walls of blood vessels, hardening them and narrowing them even more. Wow! We need to work on this message.

Others opt to say that they don't use oil because they are vegan, but oil is not of animal origin. Plus veganism is not exactly a diet. Veganism is a philosophy that asks us to avoid, in the measure possible, the use of any animal or animal product for food, clothing, or entertainment. Within the definition of veganism, there is nothing that talks about the use of oil.

Let's practice with other answers. "You see, I don't want to use oil in my food because I am avoiding fats." This sentence is clear, transmits enough information, and leaves no room for misunderstanding.

If the waitress offers another oil, butter, or margarine, the same answer can be furthered. "You see, all oil including olive oil, butter, and margarine is pure fat and I am avoiding added fats."

If later there is time and the person shows curiosity asking you why you have taken such a decision, you have the green light to go deeper into the subject, but only a bit at a time and according to the curiosity shown by the other person. Shared truths unite us.

One day the school teacher wears a sweater that spells in bold letters I AM VEGAN and a student exclaims, you are vegan!? You don't eat meat!? Well, no, I eat pure plants. They walk a few more steps and the teacher in her infinite wisdom invites with another question: "What do you understand by meat?" establishing a reason to pick up the conversation at a future time.

PART SIX

Science Is Not Neutral or Balanced, Science Is Objective

WHERE TO START

Commercial venues might not be able to change overnight.
Personally, we have superpowers.
Clean your kitchen, deny entry to anything you want to avoid.
Choose two or three recipes that entice you.
Make a list of the ingredients you will need for them.
Go shopping.
Prepare food in advance to take to work.
Save batches in the fridge.
If you make changes to 100%, you'll see results much sooner.
Keep a journal. Repeat the routine for six weeks. Do it again.

WHAT IS YOUR UNDERSTANDING OF HEALTH?

Having no detected evidence of a disease does not necessarily imply that you are healthy. Health is much more than not being sick.

Certain diseases, such as atherosclerosis, diabetes, and cancer, take a long time to develop and even more time to be detectable. The body fights sometimes for decades in a silent to and fro of promotion and regression. Think of your car with an accelerator and a brake. If by chance the brakes fail you will be sent careening. One unlucky day the immune system could lose its hold, the disease progresses, and declares itself.

I want to pause to explain something important that confuses many people. Let's say Fred quit smoking eight years ago and now bewildered is diagnosed with lung cancer. Epidemiologic statistics show that the risk for lung cancer drops immediately upon quitting, but does not approximate zero until ten to fifteen years after quitting. Maggie switched her western diet of meats, chicken, eggs, and fritters three years ago and has consumed only whole plants since then, yet a recent routine colonoscopy detected a malignant polyp. How is that possible? The risk for colon cancer is greatly reduced when you adopt a whole food plant-based diet, but there are other factors involved. Heart disease, Alzheimer's, and most other chronic non communicable diseases share this path. Let's again use the comparison of the car: if you are driving 60 miles an hour and suddenly apply the brakes, the vehicle will travel another 300 feet and perhaps even crash before coming to a full stop.

The western diet allows us to accumulate fat everywhere in our body, it increases blood viscosity, creates more free radicals, produces oxidative stress, reduces renal function, causes poor oxygenation of our tissues, and keeps us in a state of chronic inflammation. All modern diseases, including bone fragility, share the same path.

If you make a durable transition to a plant-based diet, you could reduce, avoid, or even reverse high blood pressure, excess cholesterol, type 2 diabetes, and a great many health problems plus you will lose accumulated fat. However, this will not happen without your undertaking and even so there is no guarantee.

FREEDOM

I am intrigued by those who would want to live free of responsibility. Insisting that in disease, all is inherited; that we only answer to our palate; that we can use the planet's resources as inexhaustible.

Eventually the body and the planet send us the bill. We can say that Latinos enjoy a healthy tradition, but we must defend the tradition to avoid it from falling in the hands of propaganda. Caring for what is ours is a privileged responsibility—we know that defeat is not permanent, but neither is success. Onward!

I know people who don't eat the meat of a dead animal, but drink the milk and eat the eggs of tortured animals. —Lety Ojesto

Now that you know much of what your body needs and why, you dream that your kitchen is a very special place that you can turn into an endless fountain of health. You clean it of junk. Eliminate bottled oil. You haven't bought more meat, chicken, ham, pork rinds, or sausages. No more eggs. You got used to a plant milk with no oil or sugar added. Your flours are from whole oats, corn, or wheat grain. Your spices smell fresh and fragrant. You sweeten with dates.*

*The date fruit is almost 100% carbohydrate (2 ounces = 42 g of carbohydrates; 1 g of protein; 0.3 g of fat). Dehydrated and ground a date is a great sweetener for any use.

HOW TO STOCK THE KITCHEN: AN EXAMPLE

In my kitchen each thing has its place. In one corner of the counter I have the fresh garlic plus a shallow basket of seasonal fruits: mangos, pineapple, bananas, apples, sweet potatoes, tomatoes, occasionally an avocado, and always some pumpkin.

On the long counter are dates, salt, black pepper, and a flax seed grinder.

The refrigerator door holds a jar of vegetable paste (concentrated vegetable broth), tamarind paste, mustard, olives and capers, hot sauce, tahini (sesame paste), smooth peanut butter, and a small jar of maple syrup. In the lower tier, I keep three liters of oat milk for the month, small jars of nuts (walnuts, cashews, pecans). Also, on the door is the opened bag of flax seeds and the occasional opened bag of almond meal to keep them from getting rancid because their high oil content makes them prone to oxidation.

The top shelf of the refrigerator is reserved for cut up fruits. The second and third have prepared food, such as cooked rice, beans, some soup, boiled malanga, plantain lasagna plus mushrooms in cloth bags protected from too much humidity and dehydration.

The deli drawer keeps tofu, tempeh, ginger, and sometimes homemade humus.

One of the two bottom drawers holds lemons and limes, beets, potatoes, onions, carrots, cilantro, and parsley. The other one has kale, Brussel's sprouts, bok choy, and Napa cabbage. I have two transparent boxes that sit above that bottom shelf with more vegetables: Swiss chard, spinach, leeks, parsnips, broccoli. I wrap the bottom of the cut stems in a wet paper towel to protect them from desiccation.

In the freezer, I keep a three-pound bag of blueberries and other fruits, okra, corn, edamame, and my precious homemade Puerto Rican sofrito. My compatriot Rosa taught me how to make it, then I freeze it in an ice cube tray and store the cubes in zip lock bags to start any and every rice, beans, soup, or stir fry dish.

Inside the cupboards above the counter is a revolving tray that holds the ten most used spices (cinnamon, vanilla, paprika, turmeric), nutritional yeast, vinegars, soy sauce, coffee, and bags of whole grains, such as wheat, cracked corn, and steel cut oats.

The pantry keeps non-perishables, such as rolled oats, several varieties of rice, beans, lentils, cocoa powder, cans of tomato, cans of green jackfruit, cans of lite coconut milk, cans of garbanzos and black soy beans, plus miscellaneous products like whole wheat grain flour, bicarbonate, date sugar, matcha tea and such.

I pay close attention to the environmental and social footprint of the products I bring home. For example: The cultivation of almonds requires vast amounts of water. Coffee plantations displace flora and fauna plus they especially destroy bird habitats. According to the United States Department of Labor 60% of cocoa farms employ over 1.5 million slave children. And the palm oil used in so many processed products (which are rarely in my home) is associated with deforestation.

HABITS

We are not often aware of how our habits take hold of us. Suppose you are on your way to work and you stop for coffee in a place where they also sell pastries; it smells so good and you buy a sweet bun. You tell yourself it is just a treat; however, the following week you do the same and also the following, until it becomes a routine. Letting go of this habit will require a plan.

NEW HABITS

One of the most valuable habits we can develop is preparing food in advance. Plan the day you will cook and reserve about three hours during which you will elaborate enough food for the week. Secure a dozen containers with same size lids (keeps you from wasting time looking for the right one), save the food in individual portions.

SIMPLIFY YOUR ROUTINE

Choose three recipes. Verify the ingredients you already have and make a short list of the ones missing. Grab your shopping bags and head to the market. Your mission is to come home with the ingredients for the recipes you plan to make. You might pick up a seasonal fruit or vegetable that you had not seen for a while, but in general stick to the ingredients on the list, avoid spending money unnecessarily.

Repeat week after week with small variations. Be patient. Enjoy this stage of the process. The idea is not to be entertained as much as to develop durable habits. There will be time to vary your routine.

WHAT WILL YOU EAT?

Many studies demonstrate that it is easier to make big sudden changes, rather than take small incremental steps. This might seem counterintuitive. Instead of reducing the amount of the ingredients or foods in question, consider eliminating them. It results easier to start the new path practicing it every day rather than for example only on Mondays. You might say this is extreme and that is exactly the idea, we want your system to register the change as something really different. If the new routine is too much like the previous one, the line between one and the other blurs and it will be difficult to stay focused. Besides, if you continue to tempt your taste buds with the food you intend to leave behind, you will never adjust to the new foods and flavors.

Draft your plan. Review and revise it every week. Simplify it. Above all, follow it. The more specific it is, the easier it will be to implement. Print it out and hang it in a visible place. Write down the recipes for the week and their ingredients. Indicate when you will go shopping, on what day, at what time. Include with whom you will cook. Write it on the calendar and treat it like a commitment, an appointment. Do not cancel it unless the house is on fire. As the days go by, trace your progress. Keep your eye on the target goal. You want to keep moving toward more health, one truth at a time.

SOCIAL INTERACTIONS

It is a good idea to practice showing up healthy in each and every one of the spheres that you frequent: at work, with family and friends, at religious and community events. Avoid feeling fragmented. A key word here is integrity.

Students around the Saturday table.

When you change eating habits, the world will not change with you. Old friendships will be stressed, some friends might feel rejected, perhaps they will criticize you or want you to come back into the fold, to the previous ways. Hopefully someone decides to join you, but don't hold your breath. Try to make new friends through mutual interests outside of food, like planting and keeping a native garden, cleaning a stream or minding a trail that you walk, or tending a community compost.

Try to not distance yourself from flesh and blood people. The web and computer screens are not great companies. If others are interested, consider getting together to cook simple dishes. You will develop experience, create community, and partake of some delicious fare.

This is just the beginning. You probably need to learn more. Why not become a Food For Life Instructor with the Physicians Committee for Responsible Medicine (PCRM)? You will get to know experts and advance at your own pace, receive one on one instruction in English or in Spanish, and in the end will feel more competent.

IT IS NEVER TOO LATE

Ever since I remember, but especially while in University, I practiced what I was learning with my mom, who was never in any visible hurry. She let me talk about the coral reefs, about psychology, the gods and goddesses of mythology, about butterflies and wolves. She laughed at my antics. She patiently listened to my tiresome monologues. I developed in her presence and I wonder if she was having a similar experience herself. She never bristled at my political or philosophical opinions. When I was already a physician living in the United States, I was learning too much about nutrition and the food industry, so much that I wanted to share. My visits to the island were short but intense. The family gatherings included much food I had left behind, so Jason and I always brought vegetarian dishes to share. Time went by. My mom helped interpret some of my ideas to my dad. He lived until the age of 99 eating plenty of the guanábana and mamey fruit of his youth, rice and kidney beans, some vegetables, and self-made codfish fritters he affectionately called his little poisons.

At 83 although my mom looked well, she suffered from hypertension, high cholesterol, overweight, arthritis, and depression. After my dad died, she came for a visit. When I picked her up at the airport, she was being pushed in a wheelchair. Once in the house, she walked very slowly with a cane, but true to her nature never lost her appetite. I glowed, preparing my oat, quinoa, and fruit dishes; rice with lentils; tofu scrambler with kale; broccoli with pumpkin; corn bread, and our nightly anticipated hot chocolate with soy milk. I was exceedingly happy to share with her. What a joy that she would continue to celebrate my stories. The first two days we met only at the dinner table. The rest of the time she was downstairs in her room enjoying a nap. She must have had so much stored fatigue! One evening, I sat with her through a coughing spell that interrupted her breathing. I calmed her preoccupation and mine saying she was detoxifying. The third morning she met me in the kitchen, following the aroma of fresh coffee. Already dressed, with her signature dark penciled eyebrows, she asked, "so, where are we going today?" This was the beginning of a jubilant transformation.

Her one-week visit turned into a month. I saw her marvel at the response she received from her body which hurt less with every passing day and rewarded her with more flexibility and energy. I saw her wake up to life. We walked trails, visited gardens in bloom, she loved to go window shopping. One evening we watched the animated film "Up" and filled with emotion we both remembered times gone by. When she took the plane back to Puerto Rico, she carried a journal filled with so many experiences, together with recipes we had prepared, and vivid dreams she had during those few weeks.

We spoke frequently on the phone. She told me of how she was sharing this adventure with every plumber, electrician, and gardener. She sat them down at the table with her new culinary creations and while they ate, she explained the marvel of the body and the importance of this nutrition. My heart didn't fit in my chest watching her enjoy life. Having understood simple truths, she was empowered to forge her wellbeing. My mom returned several more times. Each time with more vigor. She loved to see her reflection in the mirror, her clothes hung with elegance and she was radiant. Forgot her cane in any corner. One day she asked to ride my bicycle, something she longed from her childhood. Shaking while my hand firmly grasped the seat, she pedaled some twenty feet that for the soul were twenty leagues. The last time she returned to Puerto Rico, she was thirty-five pounds lighter and had lost ten inches around her waist. She laughed saying she could bend over to tie her shoes, her body free of inflammation and full of life. I hold those years like a treasure. If you believe you have lost too much health or that it is too late to recover it, think again.

My mom and I at the library. I inherited a thirst for learning through curiosity.

THE VIRTUAL KITCHEN TOUR

I've always been intrigued by what people buy, what they consider essential. In particular because conditions like high blood pressure, elevated cholesterol, and diabetes plus overweight, arthritis, and digestive problems are associated with their diet and perhaps they are not aware. Over so many years I asked my patients to bring diet diaries and to remember what they ate the day before. I asked in all different ways, yet none of the methods painted a complete picture with enough or reliable information.

It was then that I had the privilege to do home visits with my residents. I was given permission to enter their kitchens, look inside their refrigerators, open the freezer and pantry. The Virtual Kitchen Tour was born, an exercise that I often repeat with each patient, because it reveals someone's pattern of food like no other.

The mental tour starts with what is on top of the dinner table and the kitchen counters; it continues with the door, shelves, and drawers inside the refrigerator; we pass on to the freezer, then the pantry, and any other place where the family stores food (the garage, hallways, under the bed, in cabinets for extras).

Throughout the exercise, I write down the size of containers, the type of milk, jugs of juice, refreshments, and alcohol. I document the kind and quantity of meats, how much butter and cheese; nuts; how many dozens of eggs; candy, cookies, chips; whether they keep bottled water. A most helpful discovery was to learn that 80% of people keep one or two gallons of oil (commonly olive, vegetable, and avocado) and to know they replace it monthly. It is not rare to get reports of 5 gallons of oil. Oil is the most processed ingredient there is and the one of highest calorie density contributing 4,000 calories per pound; fat that the body has no option but to absorb and store.

The kitchen has no feelings, therefore showing some respect for the owner, we can evaluate it objectively and dispassionately without hurting anyone. I inquire about absent ingredients or those omitted during the virtual tour, such as varieties of leafy greens, sweet potatoes, pumpkins, peppers; cruciferous like cabbage, broccoli, cauliflower, collards, radishes, arugula; soy products such as edamame, tofu, and tempeh; oats, quinoa, beans and lentils; fruits, spices, nuts and seeds.

This exercise is easy, revealing, and also fun. Practice it with family and friends. The information will be very valuable, especially if you are interested in correcting imbalances in your nutrition.

THE KITCHEN ENVIRONMENT

The space where you cook should be pleasant and well lit. Consider adding a small plant easy to care for, it adds life to the place. Maintain the surfaces clean and free from street objects, mail, handbags, and keys. I suggest keeping on the counter a giant (20 inch) wooden chopping board that can accommodate all size leafy greens, vegetables, and fruits with a sharp knife always available (it invites cooking). Don't ever use the wooden board for animal meat. If you have enough room on the counter, keep the blender, food processor, and small flax seed grinder on hand. Allow fruits to ripen in some visible corner so you never let them spoil.

I keep a regular size pan, a pot, and a saucepan on the stove, clean and ready to start any dish. Organize your spices, keep them fresh. Your kitchen should be easy to use. Don't hoard stuff. Allow air to circulate between bags, boxes, cans, and bottles. Do not treat the space like a warehouse.

Next to the sink, I have a stainless steel bowl with a glass lid for the kitchen refuse that will be composted.

In a corner of the counter fruits, vegetables, and kitchen refuse for compost all share space.

HOME COMPOSTING MADE SIMPLE

In death, every animal and plant is subject to decomposition. Other animals will consume their remains giving way to insects, microbes, and earthworms to finish the process, and in the end the result will be a rich black humus. In this fertile soil, new plants will grow. They themselves will be food for other creatures, which will eventually die, and the eternal cycle of life goes on.

WHY WE SHOULD NOT THROW ORGANIC MATTER ONTO THE LANDFILL?

In the municipal landfills, 60% of the material includes garden refuse, lawn clippings, and kitchen leftovers. Deprived of air and water, surrounded by inert materials like cement, metals, and plastics, these organic residues take up valuable space for years without the opportunity to decompose.

WHAT SHOULD WE COMPOST AND HOW?

Home composting transforms our kitchen and garden leftovers into fertile soil. There are various ways one decomposes. One could bury the kitchen and garden refuse in 12 to18 inch deep trenches; we could pile them in a covered heap; or throw them inside a barrel or other aerated receptacle. In a few months we will find rich humus.

WHAT CAN WE ADD TO THE COMPOST PILE OR BIN?

The remains of fruits and vegetables: peels, leaves, corncobs, large seeds such as avocado or mango; in fact, any vegetable leftover from the kitchen plus garden refuse. If you are adding excrement, make sure it is from plant eating animals (rabbits or birds) to avoid parasites and pathogenic bacteria.

WHAT SHOULD NOT BE THROWN INTO THE COMPOST BIN?

Do not add meats, fish, or bones, because they decompose with a very unpleasant odor; you know the smell of putrid meat. Do not throw any cheese or grease, they become rancid and retard the decomposition. Do not throw the excrement of carnivores such as dogs or cats, you risk contaminating your pile with parasites. It is also not recommended that you throw flowers or small seeds, because although the core temperature of the heap may reach 140°F, they are very resistant and might germinate in undesirable places. Do not throw inside the bin any inorganic matter such as cement, glass, plastic, or metal. And of course, never throw any toxic liquid or solid inside the bin.

FROM THE KITCHEN TO THE COMPOST BIN

Keep a container by the sink where you can add the coffee grounds, tea bags, vegetable leftovers, fruit peels, eggshells (if you use eggs), spent corncobs, large seeds, and any clippings from your houseplants. A container with capacity for about 8 cups holds enough material so that you can empty it onto the outside bin once a day or every two days. Keep it covered.

BALANCE

The heap needs humidity and air, two elements absent in landfills. If the heap dries out too much, sprinkle it with some water. If it becomes too compacted, poke it with a stick and move it around to create some tunnels through which air could circulate.

TIPS AND SOLUTIONS

Too many woody branches and tangled brush would fill up the container very quickly, it is better to pile them in a corner of the yard and let them decompose at their own rate.

A receptacle filled with only kitchen refuse will be too wet and nitrogen rich. Although it will eventually decompose, it will first turn into a slimy fetid mass that will tie up your bin for too long. It is best to alternate this material with dry leaves.

A receptacle filled with only dry leaves will be very rich in carbon and although it will also decompose, you will have a dry and spongy mass without evidence of activity for way too long.

Upon removing the lid, small imbalances in the bin could greet you with an unpleasant smell or a brood of fruit flies will swarm about your face. The situation is immediately corrected by adding some dry leaves and poking holes in the heap with a long sturdy stick to improve air circulation.

If you plan to use the finished compost for an edible garden, your bin should only be filled with organic refuse. If on the other hand your plan is to fertilize ornamental plants, you don't need to be so strict.

BENEFITS OF COMPOSTING

Autumn leaves decompose on the ground contributing nutrients that nourish the next cycle of life come spring. Do not bag them. Use them to border areas of the garden. Allow the leaf litter to house caterpillars, larvae, and moths in their cocoons. Many insects sleep for seasons; birds and other small animals find nourishing food poking and scratching among the leaves. Please do not blow the leaves, you contribute to noise pollution, to air contamination, and to the destruction of pollinator habitats. The allergy season is prolonged when you are constantly blowing pollen that has already settled on the ground.

Grass clippings decompose right where they fall and enrich the soil without added fertilizers. Sadly, too many people have copied the habit of hauling precious organic material far away from their yards, only to spend money on fertilizers and weekly mowing.

If you use all the organic material for compost and recycle the rest, very little will end up as trash.

More small businesses are offering to pick up kitchen refuse and return it as finished compost. When you support them, you also benefit the environment. In the area where I live, this service is being offered by enterprising women.

If the civic association in your neighborhood is imposing antiquated aesthetic values that forbid a compost bin in the property, become an active member, educate them, be the change you want to see in the world.

Similarly, perhaps there is a tacit arrangement in the neighborhood that forces people to treat their lawn with chemicals, endocrine disruptors, pesticides, and other environmental contaminants. We make the rules, we can change them; we do this by educating each other with the truth.

An easy to care for small plant adds life to the kitchen.

EPILOGUE

We must urgently return to the earth, because that is where our roots are;
We are capable of making judicious use of technology;
Our health depends on Mother Earth and she depends on us;
We must rescue medicine from the pharmaceutical industry;
We must become independent from junk, processed, and ultra processed food;
Most of us were born with free health and we can hang on to it;
The expression of our genes is dynamic;
We enjoy better health with a better microbiota;
Health is the manifested reality of how we live and nourish our bodies, mostly;
The more plants we eat, the better is our health;
To come together is easier than to stay separated;
There is enough water, air, and food for all, if that is what we want;
Every day more people recognize this reality in the rest;
Justice is indivisible, so your struggle is my struggle.

ONE LAST TASK

Dear reader, review the topics discussed in this project, add terms to the list, try to understand them better, describe the notions in your own words, carry an intelligent conversation around them, take this book out into the world, and live it. Be a truth protector.

Design a sequence and follow a rhythm. Eliminate bottled oil, totally. Eliminate the meats, of all animals. Attack the foods you are planning to eliminate one or two per week, if necessary. Group them as you wish. Eliminate eggs. Eliminate dairy (milk and cheese). Maintain the rhythm. Do not wait too long between one step and the next. However, remember that you must substitute each ingredient you eliminate with plants that please you enough, so that they seem a fair trade.

You will be missing the salt, the sugar, the fat, because your palate was more than used to them, it was addicted to them. It is to be expected that your system will rebel. Give yourself one to three months while it adapts to less salt and grease. Perhaps you will feel irritable, get a headache, your stomach will feel bloated, and you will go to the bathroom more often. For a while you will feel tired and experience malaise. You will want to interpret all this as if you need the food you left behind; do not fall in that psychological trap. Realize that your body is going through a withdrawal and a detoxifying, both uncomfortable stretches, but self-limited. The thing that will keep you firm is your motivation, Remember it? Commit again.

All whole plants benefit us, especially when we use them instead of animal products or processed foods. Having said that, the following plant foods stand out because within their class they concentrate more nutrients, phytochemicals, and antioxidants. Many nutrition experts mention them in the top list of essential foods we should consume every day. Here they appear in the recommended amounts. It is up to you in what dishes to include them:

blueberries: one cup
ground flax seeds: two tablespoons
white mushrooms: one cup
onions and garlic: to taste
cruciferous: two cups
green tea or matcha: one to three brewed cups
soy: edamame, tempeh, tofu, two portions
beans / lentils: one cup

May it be so and turn into more health for all.

Around the table there is always room for one more.

SAMPLE RECIPES

One Portion of Oatmeal

In an uncovered pot cook 3 tablespoons of dry steel cut oats with cinnamon and enough water to cover for 20 minutes at moderate heat. Once soft, sweeten with half cup of diced fruit or berries. vary them depending on the season—strawberries, blueberries, grapes, apples, papaya…enjoy this hearty breakfast!

Baked Vegetables

In an oven proof pan, bake various vegetables all at the same time without peeling, cutting, covering, or seasoning. Heat the oven to 400° and bake for almost one hour. Try sweet potato, beet, cauliflower, carrots, pumpkin, or fennel bulb. You will have delicious nutritious food ready throughout the week.

Rainbow Salad

Start with a large bunch of leafy greens, try using Chinese cabbage, spinach, and romaine lettuce. Chop them in medium chunks and add a variety of colorful raw vegetables, grapes, oranges, ginger, cooked grains, and beans, garbanzos, lentils—whatever you like or have at hand. Keeps in the refrigerator for 3-5 days. Top it with chopped nuts and balsamic vinegar just before serving. This is a meal in itself.

Beans to Serve With Grains

Add a tablespoon of water in a pot to heat garlic and onion until tender. Turn heat to medium and add enough water to cover two cups of cubed vegetables. Once everything is soft, add two cups of cooked beans. When the beans have absorbed the vegetable flavors add a bunch of chopped leafy greens and cilantro. Turn off the heat and season. Will keep in the fridge for up to a week as. You nourish yourself and the entire family for very little money.

Bean Soup

(No oils , no beef or chicken broth). In a large pot, heat until tender minced garlic, green onions, and ginger. add sliced mushrooms, celery, habanero or red pepper, and carrot. Add cubes of potato, sweet potato, pumpkin. Top with enough water, season with salt and pepper to taste. You may use any type of cooked beans or lentils. Ten minutes before serving, mix cilantro and a bunch of chopped leafy greens, spinach, kale, or bok choy into the pot. Do not overcook.

Baked Cauliflower

Separate into florets and spread over the bottom of an ovenproof mold. Add a splash of color with red pepper strips. Drizzle a dressing of: ground cumin and coriander plus salt and pepper to 1/4 cup of lemon juice over the entire cauliflower. Bake uncovered at 400° for 45 minutes to one hour. Remove from the oven, decorate with chopped cilantro. Eat hot or cold. keeps 4 to 5 days.

Squash Soup

Turn vegetables into a creamy soup: bake one yellow pumpkin (kabocha, butternut, or calabaza) and one white or yellow sweet potato at 400° for one hour plus a handful of raw cashew nuts. Grind in a blender with plenty of water, adequately season with salt and pepper. The results are a delicious thick soup that freezes well for several months.

Burritos

This popular wrap can be filled with many delicious and nutritious well-seasoned ingredients. Burritos typically use a flour tortilla, but you can experiment with other envelopes, such as large blanched leaves of Swiss chard, napa cabbage, or collard greens.

Brussels Sprouts

Thinly slice a pound of sprouts. Sprinkle with enough salt and pepper and set aside. In a large skillet, heat 2 tablespoons of fennel seeds until fragrant. Toss all the sprouts + 3 tablespoons of water and cover immediately. Check for desired tenderness in about 5-8 minutes. Arrest cooking by transferring to a cold dish. Drizzle with white balsamic vinegar and ready!

Hummus

Typical of the middle east, this delicious dip might soon become a favorite. start with 2 cups of cooked chick peas (garbanzos). Add 1 tablespoon of sesame paste (tahini), 2 minced garlic cloves, 2 chopped green olives, one teaspoon of capers, a handful of chopped parsley, a pinch of ground cumin and the juice of 2 lemons. Mash by hand or in a food processor. Season with salt and pepper. Serve with slices of jicama, celery, carrots, broccoli, or cauliflower, instead of bread or chips.

Tofu Scrambler

Made from soy beans. Here we add colorful vegetables for an eggless scrambler. Mash a block of extra firm tofu in a warm skillet. Let most of the water evaporate. Add the diced ingredients—onion, garlic, sweet pepper, carrot, mushrooms, tomatoes, olives, capers, plus dash of turmeric, paprika, salt, pepper. last: chopped greens: cilantro, broccoli/curly kale, and 1/2 cup of green peas.

Lentil Salad

Cook one pound of lentils then drain the water.
Chop everything the size of lentils and add:
2-3 carrots.
2 celery stalks.
1 red pepper.
2-3 tablespoons fine chopped fresh dill.
2-3 tablespoons fresh tarragon.
2-3 tablespoons fresh parsley.
Dressing: 2 tablespoons of freshly squeezed lemon juice.
2 teaspoons vinegar. Salt and pepper to taste.

Lentil Burger

1 pound cooked lentils
1-2 grated carrots
2-3 mashed cooked potatoes
1/2 minced onion
2 tablespoons soy sauce
cilantro to taste
3 tablespoons ground flax seeds
Shape the patties with moist hands.
Before placing the patties on the skillet cover with mix of:
4 tablespoons nutritional yeast
4 tablespoons fine bread crumbs.
Brown on a medium hot dry skillet on one side, before flipping with spatula.

Okra

Cut each okra in half inch slices.
Place them in a dry skillet.
Add canned or fresh crushed tomato.
A drained can of chickpeas.
Season with Garam Masala spices (coriander, cumin,

cardamom, cloves, black pepper, cinnamon, and nutmeg.
Or turmeric, saffron, fennel seeds, ginger, garlic, mustard seeds, mace, star anise, tamarind, fenugreek, bay leaves.
Cover for 10 minutes until tender and serve over rice.

Chickpea Salad

1 can 15 oz chickpeas, mashed
Everything minced:
2 celery stalks; 1/4 cup orange bell pepper; 1/4 cup dill cucumber
2-3 green onions; 1/2 cup grapes; 1 garlic clove
Add:
2 teaspoons Dijon mustard; 1/2 teaspoon dry dill
1 tablespoon lemon juice
Salt and Pepper to taste

Poppyseed and Cacao Cake

From *Naked Food Magazine*, Spring 2018

2 Tablespoons poppy seeds
1/2 cup raw cashews
3 cups red kidney beans unseasoned
1 cup dates or 1/2 cup raisins
5 Tablespoons cocoa powder
2 teaspoons vanilla
1/8 teaspoon cinnamon
Creamy Layer
1 banana
1/2 cup green apple (Granny Smith)
1/2 cup raw cashews
juice of one lemon

Overnight Oats

Old fashioned (Quaker) oats
Dates chopped or mashed to a paste
baked sweet potato mashed
baked pumpkin (homemade or from a can)
mango and / or papaya
cinnamon powder
plant milk (soy, oat, almond)

Use 2-4, 6-8 cups of dry oats
plus enough fruit and liquid ingredients
to achieve a soupy consistency—keep amending as days go by
the oats will absorb liquid overnight
Method:
Mix together and refrigerate for up to 5 days.

Oat Date Cookies

Old fashioned flat oats (do not use quick or steel cut)
Dates and nuts (pecans or walnuts) equal amounts
A piece of mashed fruit for more flavor: banana, apple, strawberries.
Option of spices: cinnamon ,vanilla, chile, cardamom, ground cumin.
Soak dates in hot water or orange juice to cover for one hour
When soft, chop and mash everything together (food processor)
Add a hand full of finely chopped greens such as kale.
Form and flatten 1.5-inch balls—not too dry, not too wet
Bake on cookie sheet (parchment paper) at 375° for 20 minutes until golden
Separate from the mold while still hot and allow to cool
They keep well in a closed jar or freezer.

Baked Pears or Apples

6-8 fruits each one cut in half; Scoop out heart and seeds with a spoon
Fill with balls of chopped and mashed dates and nuts

Sprinkle with cinnamon and a few drops of vanilla
Bake a 350° for 20 to 30 minutes Delicious hot or cold

Hummus

This typical Middle Eastern dip is very versatile.
1 tsp sesame paste (tahini)
2 cups cooked garbanzos
2 cloves of garlic minced
2 olives minced
juice of 1 yellow lemon
1/4 tsp cumin
1 tsp capers
one fistful of chopped parsley
Season with enough salt and pepper.
Mash by hand or in a food processor.
Serve with celery, carrots, broccoli, cauliflower or slices of jicama, apple, bread or chips.

Mushroom Spread

Grind everything together:
12 cremini mushrooms
1/3 cup of nuts
1 tbs pine nuts
1/4 cup red kidney beans
1 tsp nutritional yeast
2 tsp bread crumbs seasoned with herbs, salt, and cayenne pepper to taste

Vegan Sushi

brown rice well cooked
rice vinegar
NORI toasted sheets (algae)

avocado slices
grated carrots
blanched asparagus
thin slices of peeled cucumber
wasabi paste
soy sauce to dip

Dinner in a Pan

Heat garlic, onion, pepper, mushrooms, tomatoes, corn, cooked beans, cooked brown rice or quinoa, cubed pumpkin, sweet potato, carrots, malanga coco, a bunch of leafy greens chopped with cilantro and chipotle, vegetable paste, diluted with water, salt and pepper to taste.

Salad Dressing

1 cup cashews soaked and drained
1 block of silken tofu
2 dates
2 cloves of garlic
2 tsp nutritional yeast
1 tsp soy sauce
2 tsp apple cider vinegar
1 cup unflavored oat milk
Combine all ingredients and blend until creamy

Sofrito Puertorriqueño

One bunch of recao (broad leaf cilantro)
2 cups sweet peppers (ajíes dulces)
One onion and one head of garlic
Grind it all and freeze in an ice cube tray, then save in ziplocks inside the freezer.
Use a cube to start rice, beans, or soups and see the difference.

MY HUMAN BODY

As I walk, talk, text, read, play, pray, write, cook, wash, sit, sleep, daydream, breathe—my body quietly lives.

My body is an amazing organism—in constant production, destruction, and repair.

It is responsible for the function of skin, muscles, bones, and nerves; renews white and red blood cells, hormones and enzymes; repairs scrapes and wounds, restores the intestinal lining; looks after my brain; gushes with tears, mucus, gastric juices, and saliva; constantly grows hair and nails; makes cholesterol, produces bile, urine, stool, and myriad chemicals.

It tries to maintain a clutter free internal environment. After digesting oats, cookies, broccoli, or burgers, it eliminates waste; it clears medicines from my system; it destroys cells that no longer work as a team or that have gone berserk. It sweeps the inside of arteries so my blood can flow freely; directs vitamins, antioxidants, and phytochemicals to vital tissues everywhere; turns inflammation on and then off on demand.

It is in constant communication with every part of itself: the feet with the hands, the thyroid with the pituitary, the pancreas with the mitochondria. It interacts with space, temperature, and surroundings. It can tell hostility from compassion and competition from group effort.

It communicates with the whole organism to let me know that I am tired, satisfied, or stressed; that I am awake; that I am hungry; that I must pay bills, do homework, go to school, work, make dinner, or that I want to make love.

My body counts on the nourishment I give it to sustain this day-to-day precious balance. This body houses myself for life.

—Ana M. Negrón MD

INDEX

Suggestions for how to use the index

If searching for a topic, first go to the Table of Contents. When you are looking for a specific term, go to the index. People and photos are listed separate at the end.whole plants

abundance, 71

adiction, 142, 252

adipocytes, 62, 97

alternate routes, 132

Alzheimer's, 56,133, 149

amino acids, 73, 124

animal protein, 37, 44, 46, 49, 69, 70, 136, 138

antibodies, 49, 58, 175

antigen, 44, 49, 175

antiinflammatory, 139, 179, 223

arthritis, 156

autoinmune, 44, 49, 139

bariatric, 112-113

bile, 45, 48, 127

Blue Zones, 54, 161

DEXA - bone densitometer, 135, 174
DHA, 79
dialysis, 125, 149
DNA desoxyrhibonucleic acid), 31, 34, 55
eagle, 199
efflux, 48, 128
elaborated, 95-98, 107, 190, 251
endothelium, 60, 88, 125-129, 191
energy, 34, 46, 65, 103, 162
environment, 98, 215, 231
enzymes, 43, 102
epidemics, 115, 205, 214, 230
epigenetic, 35-36
erectile dysfunction, 60, 131, 160
exercise, 17, 53, 135, 144
factories, 97, 189
fasting, 27, 65, 142
Food Bank, 209, 223
Food for Life Instructor, 254
freedom, 229
Fresh2You
Galileo, 70
genes, 37, 68, 137, 155, 191
glucose, 33, 45, 67, 97, 116, 133
gluten, 35, 172-173
Harvest of Empire, 214
HDL, 48, 127
healers, 166, 167
hernia, 42
Ho'Oponopono, 32
hormones, 67, 133

horse, 184
house of cards, 115, 200
infarct, 61, 126, 130, 156
influencers, 26, 219, 237
infoxication, 184, 208
joints, 52, 136, 152
ketogenic, 27, 116, 123
knee replacement, 152
laboratories, 152, 189
laboratory tests, 78, 143
leaky gut, 49
Lincoln, 199
lymph, 44
malnutrition, 181
medical education, 25, 138
medical-pharmaceutical, 19, 149
mental health, 53, 56
metabolome, 50, 51
microbiota, 79, 91, 127, 139, 173
motivational triad, 72
muscles, 52, 128
myths, 26, 106, 200, 216
nervous system, 54, 139
NIH, 24, 130
nitric oxide, 60, 125, 129, 131, 160
nitrogen balance, 73
nose, 148, 175, 207
nourishment, 102, 114, 117, 181
nutrillermo, 242
nutrition guide, 23, 82, 107, 154, 183
obesity, 62, 137, 214

Okinawa, 161, 178-179
omnívore, 78
opportunity, 118, 130, 151, 189, 199, 213
optimal, 97, 162, 243
osteoporosis, 136, 202
palate, 96, 201, 248, 252
pancreas, 43
Paradise, 189
paradox, 113, 117
PCRM, 182, 254
photography, 215
phytochemicals, 189
placenta, 32
Pleasure Trap, 142, 237
pneumonia, 40
postbiotics, 46, 102
prebiotics, 46
prevention, 43, 99, 148, 158, 204
probiotics, 46, 186
proteins, 67
psychological trauma, 152, 215
Puerto Rico, 199, 213
Raíces Global, 119
reality, 217
reductionism, 54, 88
representatives, 148, 162
review, 26, 163
rice and fruits, 122
salt, 60, 125, 189
saturated fat, 24, 37, 60, 82, 122
Saturday group, 225

screening, 143, 159
short chain fatty acids, 18, 46, 67
smoothies, 39
sofrito, 250, 272
stent, 61
subsidies, 24, 33, 115, 154
super powers, 247
supplements, 52
TMAO, 46, 127
tobacco, 53, 84, 129, 168
Trojan horse, 184
urine, 50, 73
USDA, 24, 81, 115, 181
vaccines, 167, 175
vasa vasorum, 59
vegan/ veganism, 83, 86, 243
victim, 174, 184, 204
violence, 35, 83, 159, 201, 232
virtual kitchen tour, 203, 220, 237
whole plants, 37, 102, 116, 141, 173, 178
WIC, 23, 155, 237
zygote, 32

PEOPLE

Adriana Cortés, 153, 166, 237

Alan Goldhamer, 141

Antonia Demas, 21, 203, 206

Dean Ornish, 138, 151, 153

Doug Lisle, 72, 99, 237

Caldwell Esselstyn, 61, 88, 127, 151

Evelisse Capó, 2, 166

Fabiola Flores Monsivais, 4, 21

Gloria Díaz de Puig, 215

John Robins, 21, 200

Jorge Castillo, 4, 21

Juan González, 214

LeAnne Campbell, 16

Marcela Sandoval, 3, 21, 32

Marion Nestle, 23

Myriam Masihy, 22, 28

Nathan Pritikin, 127

Rachel Carson, 199

Rosa Dávila, 209, 250

Sonia Sotomayor, 106

Saray Stancic, 3, 21, 139

Senador George McGovern, 181, 162

Sherzai, Dean y Ayesha, 133

T Colin Campbell, 62, 151, 153, 166, 204, 205

Timothy Snyder, 204, 217

PHOTOGRAPHS

1. cardiovascular system, 59

2. carbohydrates, 67

3. T Colin Campbell, 71

4. Scrabble, 74

5. musical notes, 75

6. animals for slaughter, 86

7. batteries, 92

8. flavor bombs, 96

9. The Pleasure Trap, 99

10. BMI, 112

11. early taste buds, 113

12. the resistant cells, 120

13. arteries with plaques, 126

14. circle of Willis, 132

15. class 1973, 150

16. Blue Zones, 179

17. PCRM Plate, 182

18. MyPlate.gov, 183

19. cool tiles, 196

20. Fresh2You, 209

21. intestines, 210

22. my colleague Dan, 220

23. collage, 241

24. students around the table, 253

25. at the library, 256

25. kitchen compost, 258

27. small plant, 262

bones, 51, 107, 135, 172
butchers, 186
C-peptide, 121, 232, 237, 240
CAFO, 187
calcium, 107, 129, 135
calorie, 68, 187, 211
cancer, 71, 83-84, 181, 186, 205, 248
capillares, 17, 56, 60
CARBS, 189, 190
carnivores, 107, 172, 188
cartiage, 51-52, 136
cattle industry, 24, 115, 181
cerebral infarct, 130
China Study, 70-71, 138, 178, 204
cholesterol, 126, 156, 228
chromosomes, 34
churches, 27
circle of Willis, 130
CMEs, 153
colleague, 60, 149, 177, 200-206
Comida Como Cura, 28
conflict of interests, 24, 184
consumer, 94, 184
context, 50, 88
coronary bypass, 152
coronary disease, 127, 133, 179
COVID, 114, 154, 178, 180
culture, 134
curiosity, 125, 128, 189, 191, 201-203, 233, 235
DASH, 130
demencia, 132-133

www.ingramcontent.com/pod-product-compliance
Lightning Source LLC
LaVergne TN
LVHW022156200926
848481LV00006B/1759